THE VALUE OF USING HYDROLOGICAL DATASETS FOR WATER ALLOCATION DECISIONS: EARTH OBSERVATIONS, HYDROLOGICAL MODELS, AND SEASONAL FORECASTS

ALEXANDER JOSÉ KAUNE SCHMIDT

This research was conducted under the auspices of the SENSE Research School for Socio-Economic and Natural Sciences of the Environment

The value of using hydrological datasets for water allocation decisions: earth observations, hydrological models, and seasonal forecasts

Thesis

submitted in fulfilment of the requirements of
the Academic Board of Wageningen University and
the Academic Board of the IHE Delft Institute for Water Education
for the degree of doctor
to be defended in public
on Friday, September 27th, 2019 at 3.30 p.m.
in Delft, the Netherlands

by

Alexander José Kaune Schmidt
Born in Hamburg, Germany

CRC Press/Balkema is an imprint of the Taylor & Francis Group, an informa business

Published by:
CRC Press/Balkema
Schipholweg 107C, 2316 XC, Leiden, the Netherlands
Pub.NL@taylorandfrancis.com
www.crcpress.com – www.taylorandfrancis.com
ISBN 978-0-367-42955-3 (Taylor & Francis Group)
ISBN 978-94-6395-008-4 (Wageningen University)
DOI: https://doi.org/10.18174/479159

To my loving family,

ACKNOWLEDGMENTS

Over the past five years I have been working towards the completion of this dissertation.

Back in 2014, I was working in Costa Rica after my MSc studies in IHE-Delft. Micha suggested to start a full-time PhD research, which meant that I had to come back to the Netherlands. I resigned to my job and I accepted the PhD challenge. It has been the biggest challenge until now. Now that it is over, I want to thank many people that helped and supported me along the way. Certainly, without them, this work would have been impossible to finish successfully.

I would like to deeply thank Micha Werner and Charlotte de Fraiture for offering me the opportunity to carry out this PhD research in IHE-Delft and Wageningen University. Your guidance and scientific advice was the cornerstone of this successful PhD dissertation. I am grateful for your constant support, encouragement and guidance. Micha, thank you for always motivating me to improve my work and target the level of quality that is required to obtain a PhD. More than an excellent mentor, you are truly an extraordinary person. Charlotte, thank you for your guidance, advice and patience. Thanks for your efforts to help our Land and Water group to write better papers and get our message across. Also, special thanks to Erasmo Rodríguez for offering the opportunity to do an internship in the National University in Colombia. Your guidance and scientific advice in hydrological modelling has been extremely valuable. Thanks to Poolad Karimi for offering your help since the beginning and sharing valuable references and advice in irrigation management.

Special thanks to all members of the evaluation committee. It is an honor to have all of you assessing this PhD work. Also, thanks to all the scientific experts who reviewed the publications included in this thesis.

I gratefully acknowledge the EartH2Observe project for funding this PhD research, including all the trips, conferences and publications. It has given me the opportunity to work with great and amazing experts from all around the world. Thanks to the incredible team of the Water Resources Research Group of National University of Colombia (GIREH) in Bogotá, especially to Erasmo, Inés, Albeiro, David, Nicolás, Pedro Felipe y Pedro Javier. Thanks for our scientific discussions and social events in Bogotá. Also, thanks to CSIRO Land and Water team in Canberra, Australia. Special thanks to Juan Pablo Guerschman, Marc Kirby and James Bennett for your scientific advice. I also acknowledge CONICIT/MICIT (Programa de Innovación y Capital Humano para la Competitividad, PINN) for the additional finance provided to include a case study from Costa Rica in this PhD research. Thanks for the data provided by the Instituto Meteorológico Nacional (IMN) and the Servicio Nacional de Aguas Subterráneas, Riego y Avenamiento (SENARA) de Costa Rica. Also, special thanks to CeNAT (PRIAS and Observatorio Climático) in Costa Rica for the internship position and information exchange.

The contribution of a number of colleagues of IHE-Delft, Vrije Universiteit Amsterdam and Deltares was crucial for the completion of this work. Special thanks to Clara Linés, Gonzalo Peña, Juan Carlos Chacón, Gabriela Cuadrado, Patricia López, Anouk Gevaert and Ted Veldkamp. Ted and Anouk, thanks for our conversations and scientific discussions about our paper, it certainly motivated me to work better, and be more organized. Clara, thanks for your support and help, especially at the beginning of our PhD when we were trying to figure out what to do. Gonzalo and Juan Carlos, you have been friends in IHE since the beginning, thanks for your coding advice, you saved me more than once. Gaby, thanks for our conversations about Costa Rica and our paper idea and development. It was certainly a new challenge during the last stages of my PhD, which also contributed to my writing. Patricia, I cannot thank you enough. Our meetings in IHE were crucial, at a time when I had many ideas, but I had no clue if they made any sense. Thanks for listening and asking the right questions. At the end the "value of additional information" found its way ☺

The last year of my PhD, I started to work at a research and consultancy company called FutureWater. Starting to work again, helped me to be more organized and more efficient, both at work and in my PhD. Special thanks to my colleagues Sonu, Jonna, Martijn, Gijs, Peter, Arthur, Corjan, Johannes, Raed and Sergio.

To the endless colleagues and friends I met in Delft. Many of you are still around, others have gone with the years. I deeply value your friendship, and advice. Conversations and discussions about life, culture, political views, religion, football or just relaxing with friends and colleges, has made this part of my life pretty unique and memorable. Friends are many, and possibly I am leaving more than one out of this list. Thanks to Mohan, Mohaned, Omar, Yared, JuanCa, Neiler, Maurizzio, Aki, Gaby, Francesco, Alessandro, Maria, Musaed, Alida, Mario, Vitali, Eiman, Jonatan, Shahnoor, Thaine, Gonzalo, Miguel, Sida, Clara, Nadine, Abdi, Chris, Polpat, Quan, Tarn, Natalia, Sofia, Maria, Mauricio, Mary, Iosif and Jeewa, and so many others. In Colombia, special thanks to Gonzalo, a truly exceptional friend, hope to see you again soon! In Bolivia, special thanks to Angi, a loyal friend since the MSc studies, gracias totales! In Brasil, special thanks to Chris, a truly kind heart. Thanks for our coffees in IHE, which were needed to escape the routine. Un fuerte abrazo! Special thanks to Cesar and Adriana, who helped me unconditionally during my stay in Bogotá and in other endeavors in my PhD. Special thanks also to Benno. Talking with you gave me another perspective of reality outside the PhD life, which was needed and important to go forward. Thanks for all your advice.

My friends, "the gang" from high school and university back in Costa Rica always had something to say and we stay in contact after so many years. A special acknowledgment to them, because I missed many activities and special occasions due to the distance, and still today I can count on them and they count on me for sharing our challenges and successes. The 300 WhatsApp messages when I wake up in the mornings were always giving me motivation to start the day ☺. Special thanks to "the gang" Nela, Migue, Albert, JuanCa, Julio, Ale, Ponchi, Dani, Iker and Adrian. Nela thank you for always listening, you have the biggest heart. Un abrazo gigante!

Mami y papi, hermanos. What a family! A truly example of unity, hard work and success. I love them all very deeply. My brothers Andrés, Daniel and Marcel, you are part of my life. We lived so many years together, sharing things, partying and playing sports, but then also so many years separated. In the last years Andrés got married and had a beautiful daughter. I hope we can reunite and share many nice moments again in the future. To my father, you have been my role model and inspiration. Thanks to your unconditional support, I am here now receiving this

PhD degree. To my mother, you have always been there for me. Always listening, caring and giving advice, but also knowing when to give us space for our benefit. You are an essential part of my success. I love you all!

Dear Audry, we met at the beginning of my PhD, and during this time our relationship grew stronger, we will get married soon and start our own family. Nevertheless, during my PhD, you had to deal with my stresses and sometimes bad humor. It was only because of your patience and support that we managed to go forward and grow as a couple. Thank you so much for all your kindness and love, and for all the adventures we have together. Now that this PhD is finished, I want to focus more on our plans and dreams. Te amo!

Alex Kaune

SUMMARY

Hydrological data and information on the availability of water are essential to support water allocation decisions, especially under increasingly water scarce conditions. However, water allocation decisions are often taken based on uncertain hydrological information, which may lead to agricultural production loss and possibly impact on local livelihoods. This study aims to evaluate the availability of hydrological data and information at different scales, and assess the value of using additional hydrological data and information in enhancing water allocation decisions, especially in irrigated agriculture.

An index tool was developed thatmeasures the quality of hydrological information in irrigated agriculture. The index is calculated based on a compound that considers the period of record, temporal and spatial resolution of the data available used in planning and operation of large irrigation districts. In these irrigation schemes, the observed discharge in water extraction sites, supply in the main canals, and irrigation demand estimates were evaluated to determine the rate of occurrence of sub-optimal water allocation decisions. Through this method, excellent index results were found for an irrigation district in Australia (Murrumbidgee district), while irrigation districts in Colombia (Coello district) and Costa Rica (DRAT district) showed fair to poor information quality, which correspond to a higher rate of occurrence of sub-optimal water allocation decisions. The results imply that the use of additional hydrological information is beneficial in reducing the rate of occurrence of sub-optimal water allocation decisions, ultimately contributing to reduce agricultural production loss.

The challenge is to improve water resources management in irrigated agriculture using additional hydrological information from global and local datasets, including satellite, hydrological models and reanalysis datasets. For this matter, a hydro-economic framework was developed to value the use of these hydrological datasets in river discharge estimates for irrigation area planning. The risky outcomes in terms of annual agricultural production due to the actual frequency of occurrence of water scarcity were generated for a reference and planned irrigation areas, resulting in a Relative Utility Value (RUV) that expresses the utility of the information used. Based on this framework, results showed that using an ensemble of global hydrological models in Australia provides more robust estimates in surface water availability compared to when using any of the single global models that constitutes the ensemble. A comparison of the information content in the ensemble shows that an ensemble with a period of record of fifteen years has an information content equivalent to a single model of thirty years.

Moreover, in Colombia, a reference area that can be irrigated is established using thirty years of observed river discharge data. Areas are then obtained using simulated river discharges from six local hydrological models forced with global precipitation datasets (CHIRPS and MSWEP), each calibrated independently with a sample of five years extracted from the full thirty year record. The utility of establishing the irrigated area based on simulated river discharge simulations is compared against the reference area through a pooled Relative Utility Value ($PRUV$). Results show that for all river discharge simulations the benefit of choosing the irrigated area based on the thirty years simulated data is higher compared to using only five years observed discharge data, as the uncertainty of *PRUV* using thirty years is smaller. Hence, it is more beneficial to calibrate a hydrological model using five years of observed river discharge and then extending it with global precipitation data of thirty years as this weighs up against the model uncertainty of the model calibration.

Conservative estimates of the available resource in rivers and reservoirs may lead to allocated water to be underestimated due to conservative estimates of future (in) flows. Water allocations may be revised as the season progresses, though inconsistency in allocation is undesirable to farmers and may lead to an opportunity cost in agricultural production. We assess the benefit of using reservoir inflow estimates derived from seasonal forecast datasets to improve water allocation decisions. A feedback loop between simulated reservoir storage and emulated water allocations for General Security (allocated to irrigating) was developed to evaluate two reservoir inflow datasets (Poama and ESP) derived from the Forecast Guided Stochastic Scenarios (FoGSS), a 12 month seasonal ensemble forecast in Australia. We evaluate the approach in the Murrumbidgee basin, comparing water allocations obtained with an expected reservoir inflow from FoGSS against the allocations obtained with an expected reservoir inflow from a conservative climatological estimate (as currently used by the basin authority), as well as against those obtained using observed inflows (perfect information). The inconsistency in allocated water is evaluated by determining the total changes in allocated water made every 15 days from the initial allocation at the start of the water year to the end of the irrigation season, including both downward and upward revisions of allocations. Results show that the inconsistency due to upward revisions in allocated water is lower when using the forecast datasets (Poama and ESP) compared to the conservative inflow estimates (reference) which is beneficial to the planning of cropping areas by farmers. Over confidence can, however, lead to an increase in undesirable downward revisions. This is more evident for dry years than for wet years. Even though biases are found in inflow predictions, the accuracy of the available water estimates using the forecast ensemble improves progressively during the water year, especially one and a half months before the start of the cropping season in November.

CONTENTS

1

GENERAL INTRODUCTION

1.1 WATER SCARCITY AND IRRIGATED AGRICULTURE

Water scarcity occurs when available water cannot satisfy water demands for drinking, sanitation and food production. With increasing global population and a changing climate water scarcity situations are occurring more often (Rijsberman, 2006). Locally, either at country or basin level, water can be physically scarce due to low precipitation and high temperatures from climate variability, exacerbated by droughts and climate change (Van Loon and Lanen, 2013; Van Loon et al., 2016). However, the impact of climate extremes depends on local economic and social conditions, which vary considerably among countries (Naumann et al., 2014). Water scarcity situations occur not only due climate conditions, but also due the lack of investments in infrastructure and water management strategies at planning and operational stages (FAO, 2013; Loon and Lanen, 2013; Seckler et al., 1998). Understanding the balance between water availability and demand is crucial to safely manage the water resource and avoid economic, environmental and social losses. In the water balance, irrigated agriculture accounts for 70% of the world's total freshwater withdrawals (FAO, 2013). Since the Green Revolution during the 1960s, the use of water for irrigation has been justified to increase crop yields and satisfy growing food and fibre demands for a growing population (FAO, 2016). Thanks to the development of irrigation infrastructure combined with the use of agrochemicals, and high yielding crop varieties, irrigated agriculture using only 18% of world's arable land is responsible for 40% of the total food production (Schultz et al., 2009).

Due to irrigation, farmers use less land while producing more food. However, water for irrigation has significantly altered the hydrological and environmental characteristic of many basins (Kirby et al., 2014a; Restrepo and Kettner, 2012). Irrigation managers must often justify the use, efficiency and productivity of water in competition and comparison with other uses and users. The environmental impact on water quality, land degradation and salinization due to unsustainable agricultural development and the contribution of agriculture to climate change (e.g. greenhouse gas, fertilizers, fossil fuel) is reason for concern, both for the environment and the agricultural sector. In addition, climate change itself poses a challenge in mitigating impacts and finding adaption measures in land and water development. In the post-Green Revolution era the challenge is to find new methods and strategies in the agricultural sector. Research and investments have focused in using information technology, precision agriculture, and environmental friendly techniques to pursue sustainable water use and agricultural development (Baumüller, 2018; Chuchra, 2016; Far and Rezaei-Moghaddam, 2018; King, 2017; Nikouei et al., 2012; Pareeth et al., 2019). Such a trend has been supported by several interlinked goals in the 2030 Agenda for Sustainable Development, in particular Sustainable Development Goal (SDG) 6: "Ensure availability and sustainable water management"; SDG 2: "End hunger, achieve food security, improve nutrition and promote sustainable agriculture" and SDG 13: "Take action to combat climate change and its impacts".

1.2 WATER ALLOCATION DECISIONS IN IRRIGATION

Irrigation infrastructure supply water to the crop fields (Malano and Hofwegen, 1999) through main, secondary and/or tertiary canals or conduits. Typically, the canal system is operated by a user's association, public or private entity. Many of these schemes were built in the 1960s, with the intention to boost local agricultural production and secure food security (FAO, 2016). Today, large irrigation systems provide key food and fibre production for local economies and trade.

In many irrigation schemes the performance of service delivery is low to mediocre (FAO, 2007). There is a critical need for improvement in irrigation management, particularly in those areas where water scarcity affects agricultural productivity (Richter, 2014; Rijsberman, 2006). Arid and semi-arid zones regularly face water scarcity in agriculture (e.g. Spain and Ethiopia) but also more humid tropical regions (e.g. Costa Rica and Colombia) are affected due to inter-annual variability (Córdoba-Machado et al., 2015) and climate change, causing unexpected production loss. Agricultural livelihoods depend on proper water allocation decisions to overcome service deficits and water scarcity.

Allocating water is the process of sharing the available water among users (Hellegers and Leflaive, 2015; Le Quesne et al., 2007). Social, economic and political dimensions influence the way that water is allocated, which may influence the distribution of wealth among water users (Hellegers and Leflaive, 2015). Established rules, regulations and water policies support the water allocation process between competing water users. In countries such as Australia and United States, the water allocation process is supported by clear water policy and regulations designed to minimize economic losses due to water scarcity. However, major agricultural production losses and unbalances in the environment occurred due to extended drought periods (Richter, 2014). Water allocation rules had to be changed several times during the last years, due to incompatibility in water scarce situations. For every water allocation reform, water volumes for irrigation remain high and often controversial due to environmental demands (Bark et al., 2014).

In an increasingly water scarce world reliable information on available water is key to allocate suitable amounts and avoid undesired socio-economic impacts, conflicts and reforms. Reliable information and adequate tools are required to support water allocation decisions (Hellegers and Leflaive, 2015). Frameworks have been developed to assess and predict water allocation decisions at basin and irrigation district scale. For example, at basin scale Dai and Li (2013) developed an allocation model for planning agricultural water management focused on optimizing water and land use. At irrigation system scale, integrated allocation models have been developed to consider economic variables (crop prices and yields) for water allocation strategies (Kim and Kaluarachchi, 2016). Also, coupled agronomic-economic models have been applied in irrigation systems to assess water use in upstream and downstream farming plots (Ben-Gal et al., 2013). Even artificial intelligence techniques are used for a systematic water allocation scheme to mitigate drought threats (Chang and Wang, 2013).

1.3 THE NEED FOR GOOD QUALITY AND ACCESIBLE HYDRO-METEOROLOGICAL DATA

Hydrological data provide a quantifiably form of information that can be stored, processed and made available for the general public, water professionals, and specific decision makers in basin authorities and irrigation systems. Ideally, hydrological data are collected from measurements based on an optimal station network design (Chacon-Hurtado et al., 2017), with specific temporal and spatial requirements to support water allocation decisions. These measurements provide data with a standardized quality (WMO, 2008), depending on the method of measurement, measurement uncertainty, station density, frequency of measurement and length of records. Interpolation methods may be applied to obtain gridded hydro-meteorological datasets from these station measurements and simulate the surface water availability (Raimonet et al., 2017). Sometimes, hydro-meteorological data are available from measurements conducted as part of ad hoc local studies. However, the lack of standardized procedures reduces the full potential in its use for water resources assessments.

Where measuring networks are not available, governments are faced with the dilemma of investing in hydro-meteorological data and information systems. For example, in the Netherlands, back in the 1950s, the government was struggling to take the decision of developing a large hydraulic project (Delta Works), which would include a measurement and monitoring systems to protect and inform citizens from storm surges and floods. At the time, the economic cost of this project was high, and the government did not invest in it, which finally led to the known catastrophic flood event in 1953 (Gerritsen, 2005). The aftermath of the event sparkled decision makers to finally implement the project. Today, a complete hydro-meteorological data and information platform is available that records data of hydro-meteorological variables which are used to support decision making in the water sector (Gerritsen, 2005; Rijkswaterstaat, 2018). Similarly, in other countries the development or improvement of hydro-meteorological data and information systems may require large economic investments. In Costa Rica, for example, 1.5 million US dollars are required to maintain preserve and modernise the hydro-meteorological network (MINAE, 2008).

The existence of well-functioning measuring networks does not guarantee information will be used for better decision making. Sometimes data are not easily accessible because they are scattered over different departments and local politics prevent enabling environment for data sharing. In other cases practical issues such as the lack of online platforms or standardized formats for sharing hamper easy access. Even where good quality data and information tools are available and easily accessible, policies and regulations may render its use in water allocation decision sub-optimal. For example, from 2001 to 2009 one of the worst droughts occurred in Australian modern history (Millennium Drought), reducing the water availability for human, environmental and agricultural consumption (Richter, 2014). Agricultural crops were lost and farmers struggled to maintain their investments due to water scarcity. At the time, complete hydro-meteorological datasets and information about water resources were available from the Bureau of Meteorology (BoM) and other regional institutions. However, inadequate water policies and regulations failed to prevent the socio-economic and environmental impacts and led to unexpected losses in agricultural production (van Dijk et al., 2013b).

1.4 THE VALUE OF USING HYDRO-METEOROLOGICAL INFORMATION

Investing in information can be valuable if that information leads to a benefit or a reduction in costs or losses as a result of better decisions (Hirshleifer and Riley, 1979). Three factors are considered in order to determine the value of information: 1) decision maker's beliefs; 2) the costs linked with decisions; and, 3) the response in taking actions based on new or additional information. If current available information is derived from uncertain estimates, then additional information can provide a benefit and enhance decision making (Bouma et al., 2009; Macauley, 2006). With perfect available information no added value is expected in the decision process because the existing information already provides a clear choice of alternatives. Additional information could have a high marginal value depending on the existing availability and quality of information.

Several studies have assessed the value of additional hydro-meteorological information from monitoring networks (Alfonso and Price, 2012; Malings and Pozzi, 2016), from drought and flood predictions (Quiroga et al., 2011; Verkade and Werner, 2011) and space derived weather and water quality monitoring (Bouma et al., 2009; Macauley, 2006). The typical study for evaluating the value of weather information in agriculture compares expected production gains using climatological information against gains using forecasted rainfall (Cerdá and Quiroga, 2011; Luseno et al., 2003).

The longer the record of hydro-meteorological measurments the more the data are reliable for assessing water availability and supporting water allocation decisions. However, in practical terms stakeholders cannot wait for thirty years to obtain a complete period of record. In addition, investments in data networks are costly. A possible cost-effective solution is combining (incomplete) local data with a complete time period (over 30 years) from global earth observations and models (Beck et al., 2017a, 2017b; Funk et al., 2015; Schellekens et al., 2017). Additional hydrological information from earth observations and hydrological models may be a useful source to complement the local datasets and improve water management at basin and irrigation district scale. These global datasets are available as a grid format for the entire planet with a combination of local, reanalysis and satellite observation datasets. The resolution and estimations of these products have been enhanced under continuous research and testing (Beck et al., 2017a, 2017b, 2018; Schellekens et al., 2017), and its application in local conditions has been applied by researchers and water professionals to overcome local data scarcity (López López et al., 2016, 2017; Rodriguez et al., 2017).

1.5 PROBLEM STATEMENT

Available local hydrological data may be limited, not provided at the adequate time period and resolution. Hence, available hydrological data may be of low quality for water allocation decisions. In addition, depending on country policies and regulations, the access to hydrological data may be restricted. The lack of quality and accessibility of hydrological data can lead to inaccurate estimates in climate variability, poor water availability assessments, and sub-optimal decisions regarding the supply and allocation of the water in irrigated agriculture. In countries facing the combined challenges of water scarcity and pressure to provide for rural farming livelihoods, better information may lead to better outcomes. While studies on the value of information for water allocation decisions are available, none of them is specifically geared towards water allocation decisions in irrigated agriculture. In irrigated agriculture, different types of water allocation decisions are made. For example, water authorities or irrigation developers need to decide the size of new irrigation areas, taking into account possible investments in pressurized systems, canals, and hydraulic structures. At the start of each cropping season decisions about which crop to grow are relevant. At the short term operational stage, daily decisions related to opening and closing control gates and supplying water are key.

Water policy and regulations include specifications about the use of hydrological data and information. However, in some countries or regions, these specifications lack guidelines and methods to evaluate the quality of the available hydrological datasets and its adequate application at different scales. In addition, the use of earth observations is limited and its potential value has not been evaluated. Regions with reduced in-situ measurements may benefit from global datasets as the water availability can be better determined for planning purposes in agriculture, urban areas and the environment.

This PhD research contributes to the science available about the value of data and information and aims to assess the benefit of using hydrological datasets in supporting water allocation decisions (especially in irrigated agriculture), implementing state of the art in hydro-economic modelling, and adopting and adapting available methods to establish water availability and demand. This research assesses the benefit of combining locally available - but often incomplete datasets - with global earth observation products.

1.6 RESEARCH OBJECTIVES

The main objective of this research is to contribute to the development of new tools and methods in valuing hydrological data and information and assess if additional data can enhance water allocation decisions in irrigated agriculture.

The following are the specific research objectives

1. Develop a tool for assessing the availability of hydrological information in large irrigation districts and evaluate water allocation decisions being taken (Chapter 2).

2. Assesses the benefit of using an ensemble of global hydrological models to improve surface water availability estimates in irrigation area planning (Chapter 3).

3. Assesses the benefit of using global precipitation datasets to improve surface water availability estimates in irrigation area planning (Chapter 4).

4. Assesses the benefit of using seasonal streamflow forecasts for water allocation decisions (Chapter 5).

This research was carried out in the framework of the EartH2Observe project (Global Earth Observation for Integrated Water Resource Assessment, http://www.earth2observe.eu/). EartH2Observe was funded under the DG Research FP7 programme of the European Union with a total duration of 4 years, from January 2014 until the end of 2017. The EartH2Observe consortium was composed of 27 partners and 4 associate partners, 23 from the EU and 8 from non-EU. The main aim of the project was to contribute to the assessment of global water resources through the development and application of state-of-the-art earth observations, reanalysis datasets and hydrological models. Hence, allowing improved estimates of the available water in order to support water management decisions at global and local scale. Tests were conducted in key basins within a variety of countries with different water policies, data availability and climate conditions such as: Colombia (Magdalena-Cauca basin), Australia (Murray-Darling basin), Spain (Ebro basin), Ethiopia (Blue-Nile basin), Morocco (Oum Er Rbia basin) and Bangladesh (Brahmaputra basin). This dissertation focuses on three case studies within the EartH2Observe project where irrigated agriculture plays a major role: In Australia, the Murrumbidgee Irrigation Area has been called the food bowl of New South Wales and extends over an area of 150,000 hectares with multiple crops including grapes and rice (MI, 2015b). In Colombia, the Coello Irrigation district located three hours from the major city of the country, provides land and water for rice, maize and cotton production (Garces-Restrepo et al., 2007; Urrutia-Cobo, 2006). In addition, in Costa Rica, the DRAT irrigation district provides key rice and sugar cane production for satisfying major rice consumption and contributing to derived food products (CONARROZ, 2015; SENARA, 2014).

1.7 OUTLINE OF THE THESIS

This thesis is structured in seven chapters. Chapter 1 contains the general introduction and Chapter 6 contains the synthesis, and recommendations for future work. Chapters 2 to 5 have been published or are in the process of publication. The sequence of chapters is based on the development of frameworks.

Chapter 2: A tool to assess available hydrological information and water allocation decisions

Chapter 3: The benefit of using an ensemble of global hydrological models in surface water availability

Chapter 4: Can global precipitation datasets benefit the estimation of the area to be cropped?

Chapter 5: The benefit of using an ensemble of seasonal streamflow forecasts for water allocation decisions

2

A TOOL TO ASSESS AVAILABLE HYDROLOGICAL INFORMATION AND WATER ALLOCATION DECISIONS

Based on: *Kaune, A., Werner, M., Rodríguez, E., Karimi P., de Fraiture C., 2017. A novel tool to assess available hydrological information and the occurrence of sub-optimal water allocation decisions in large irrigation districts. Agric. Water Manag. 191, 229-238. https://doi.org/10.1016/j.agwat.2017.06.013.*

Abstract

Hydrological information on water availability and demand is vital for sound water allocation decisions in irrigation districts, particularly in times of water scarcity. However, water allocation decisions are often taken based on uncertain hydrological information, which may lead to sub-optimal decisions and agricultural production loss. This study aims to assess the availability of hydrological information in large irrigated areas (>250 km²) and evaluate water allocation decisions being taken. An index tool that measures the level of availability of hydrological information in irrigation districts that is used in planning and operation was developed. The index is calculated based on a compound that considers the period of record, temporal and spatial resolution of the data. Contingency tables that compare the observed discharge in water extraction sites, supply in the main canals, and irrigation demand estimates, were generated allowing the rate of occurrence of sub-optimal water allocation decisions to be determined. Through this method, excellent index results were found for an irrigation district in Australia (Murrumbidgee district), while irrigation districts in Colombia (Coello district) and Costa Rica (DRAT district) showed fair to poor information availability, which correspond to a higher rate of occurrence of sub-optimal water allocation decisions. The results imply that the use of additional hydrological information is beneficial in reducing the rate of occurrence of sub-optimal water allocation decisions, ultimately contributing to higher crop yields.

2.1 INTRODUCTION

Irrigated agriculture accounts for nearly 70% of the world's total freshwater withdrawals (FAO, 2011) and over the years has significantly altered hydrological conditions in streams (Al-Faraj and Scholz, 2014; Jiang et al., 2015; Kirby et al., 2014b). Even though management strategies have been developed to overcome water limitations and provide reliable and stable water supply for agricultural production, droughts have in cases led to high impacts on production and livelihoods (Connor et al., 2014; Richter, 2014). Given the pressing water scarcity issue in agricultural development (de Fraiture and Wichelns, 2010; Rijsberman, 2006), it is crucial to take informed water allocation decisions, especially in irrigation districts with high water consumption, such as large surface canal systems, where users rely on the water supply system for agricultural production to guarantee economic welfare and food security (Malano and Hofwegen, 1999).

As water becomes scarce, efficient decision-making based on solid information becomes increasingly important. In reality, water allocation decisions are often based on incomplete or uncertain hydrological information (Svendsen, 2005). Poorly informed water allocation decisions can lead to agricultural production loss and possibly impact on local livelihoods. Assessing the available hydrological information and the potential improvement due to additional information can be of considerable value to irrigation districts.

The value of information theory, developed in the field of economics (Hirshleifer and Riley, 1979), considers three factors in order to determine to what extent additional information has, or does not have a value; the beliefs of decision makers; the costs associated with decisions; and, the response in taking actions in light of new information. Additional information can add value and improve decision making, but only if currently available information is uncertain (Bouma et al., 2009; Macauley, 2006). This implies that, if the available information is perfect, then no value is foreseen in the decision process because the current information already leads to a clear choice of alternatives. If there is no or little information available, then the additional information would have a high marginal value.

Previous studies that determined the marginal value of additional information (Alfonso and Price, 2012; Cerdá and Quiroga, 2011; Macauley, 2006; Quiroga et al., 2011) focused on monitoring networks, irrigation frequency and weather information for agricultural production and management (including damage and loss). Even though several authors have established the marginal value of additional information for agricultural purposes, a gap exists in determining how hydrological information can be evaluated for a particular user within a river basin, such as in irrigation districts.

This research hypothesizes that if decision makers in the irrigation sector have the right tools to assess the available hydrological information, they will invest more in hydrological information when it makes economic sense. The aim of this chapter is to develop a tool for assessing the availability of hydrological information in large irrigation districts and evaluate water allocation decisions being taken.

2.2 METHODS

2.2.1 Establishing hydrological information requirements for water allocation

The allocation of water resources requires different types of decisions. In the long term planning phase, these include determining the viability of developing new irrigation districts or extending existing districts, which may entail the construction of canals and hydraulic structures for supplying additional demand (FAO, 2007). In the medium term, these include the types and extent of crops to be planted given the expected available water. In the short term, these include daily decisions related to opening and closing control gates and supplying water to sectors and tertiary units (FAO, 2007). These decisions are supported by available data and information and governed by water policies, regulations and guidelines (Australian government, 2008; MI, 2013a; MINAE, 2009; MINAGRICULTURA and INAT, 1997; MinAmbiente, 2014; SENARA, 2014).

Even though the water allocation processes can vary between irrigation districts, the end goal is similar, namely providing a service of water supply (Malano and Hofwegen, 1999). Determining the water availability and demand for irrigation is key information for providing this service. In this study, the hydrological information requirements are defined as the priority ground measurements and data required for determining the water availability and irrigation demand in the long term, medium term and short term phases.

For the long term planning phase, variables such as precipitation, temperature (for evapotranspiration estimates), river discharge (runoff), water storage (including groundwater and surface water reservoirs), cropping patterns and soil characteristics (including texture and effective depth) were established as hydrological information requirements (Figure 2.1). For the medium term planning and short term operation the required hydrological information was established in discharge control locations, including discharge information at water extraction sites, water supply at intakes (main, secondary, tertiary canals and user intakes) and irrigation demand in the fields (Figure 2.1).

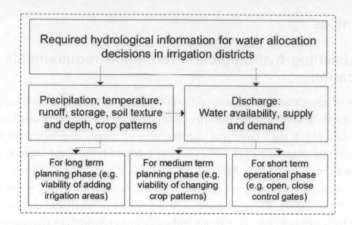

Figure 2.1. Hydrological information requirements for water allocation in planning and operational phases in irrigation districts.

2.2.2 Developing the hydrological information availability index

The hydrological information availability index is a novel approach that tries to establish the available hydrological information in large irrigation districts through a scoring method.

The availability of hydrological and meteorological data was evaluated according to the spatial and temporal resolution and the period of record, through scoring each aspect on an ordinary scale between 5 and 1 where 5=excellent; 4=good; 3=fair; 2=limited; 1=poor. If a measurement is not available the score was set to zero.

For the long term planning phase, the information availability of river discharge and storage in the basins from which the irrigation water is extracted was evaluated, while the remaining variables are evaluated within the limits of the irrigation district. The spatial resolution of river discharge, precipitation, and temperature were determined based on the station inverse density (Table 1). The spatial resolution of crop patterns and soil characteristics are defined as the percentage of area in the district with available information. For the spatial resolution of river discharge, the lowest score (1) is assigned to a spatial representativeness of 1000 km²/station, a rate set by WMO for mountainous basins (WMO, 2008). The highest score (5) was assigned to a spatial representativeness lower or equal than 400 km²/station, assuming at least five river discharge stations in a basin of close to 2000 km². For the spatial resolution of precipitation and temperature stations, the lowest score (1) is assigned to districts with only one station in the entire command area. Spatial representativeness lower or equal to 30 km²/station were assigned the highest score (5) for enhanced water demand estimation for irrigation districts larger than 250 km² (Table 2.1).

Table 2.1. Scoring method for hydro-meteorological variable i required for long term planning phase in irrigation districts. Precipitation, temperature, crop patterns, soil texture and depth evaluation in the limits of the irrigation district. River discharge and storage evaluation in the basins from which the irrigation water is extracted.

Independent score a_{r_i}, a_{t_i}, a_{s_i}	Period of record, r_i (years) Precipitation, temperature, river discharge and storage	Temporal resolution, t_i (month) Precipitation, temperature, river discharge and storage	Spatial resolution, S_i (km²/station) Precipitation and temperature	Spatial resolution, S_i (km²/station) River discharge	Spatial resolution, S_i (%) Soil texture depth, crop patterns
5 (Excellent)	$r_i \geq 30$	$t_i < 1$	$S_i \leq 30$	$S_i \leq 400$	$S_i = 100$
4 (Good)	$20 \leq r_i < 30$	$t_i = 1$	$30 < S_i \leq 65$	$400 < S_i \leq 600$	$80 \leq S_i < 100$
3 (Fair)	$10 \leq r_i < 20$	$1 < t_i < 12$	$65 < S_i \leq 125$	$600 < S_i \leq 800$	$50 \leq S_i < 80$
2 (Limited)	$5 \leq r_i < 10$	$t_i = 12$	$125 < S_i \leq 250$	$800 < S_i \leq 1000$	$30 \leq S_i < 50$
1 (Poor)	$r_i < 5$	$t_i > 12$	$S_i > 250$	$S_i > 1000$	$S_i < 30$

Table 2.2. Scoring method for discharge information in established control locations j in irrigation districts (medium term planning phase). Evaluation in water extraction sites; hydraulic intakes (main, secondary, tertiary canals and user intakes) and irrigation demand in the fields.

Independent score a_{r_j}, a_{t_j}, a_{s_j}	Period of record, r_j (years)	Temporal resolution, t_j (day)	Spatial resolution, S_j (%)
5 (Excellent)	$r_j \geq 30$	$t_j < 1$	$S_j = 100$
4 (Good)	$20 \leq r_j < 30$	$t_j = 1$	$80 \leq S_j < 100$
3 (Fair)	$10 \leq r_j < 20$	$1 < t_j < 1\text{month}$	$50 \leq S_j < 80$
2 (Limited)	$5 \leq r_j < 10$	$t_j = 1\text{month}$	$30 \leq S_j < 50$
1 (Poor)	$r_j < 5$	$t_j > 1\text{month}$	$S_j < 30$

Table 2.3.   Scoring method for discharge information in established control locations j in irrigation districts (short term operational phase). Evaluation in water extraction sites; hydraulic intakes (main, secondary, tertiary canals and user intakes) and irrigation demand in the fields.

Independent score a_{t_j}, a_{s_j}	Temporal resolution, t_j (hour)	Spatial resolution, S_j (%)
5 (Excellent)	$t_j < 1$	$S_j = 100$
4 (Good)	$t_j = 1$	$80 \leq S_j < 100$
3 (Fair)	$1 < t_j < 24$	$50 \leq S_j < 80$
2 (Limited)	$t_j = 24$	$30 \leq S_j < 50$
1 (Poor)	$t_j > 24$	$S_j < 30$

For the medium term planning and short term operational phase, the score for the spatial resolution was established considering the percentage of control locations (e.g. intakes of secondary canals) in the district with available discharge information (Table 2.2, Table 2.3).

For the medium term, the period of record of that information is crucial for adequate planning, as past discharge measurements can be used to establish statistics and trends, and recommend water allocation for the next season.

For the short term, operational phase temporal resolution ranges from one hour to 24 hours (Table 2.3). A coarser temporal resolution (between one day and one month) was established for the medium term planning phase (Table 2.2) because decisions on planning the next crop season are to be taken, and from one month to one year for the long term planning phase (Table 1). Information on the period of record was included in both the medium term and longer term phases, where 30 years of historic data was assigned as excellent information availability (score 5), due to its usefulness in assessing climate variability. For the short term operational phase, the period of record was not included as this is less relevant for making day-to-day decisions.

In addition, a weighting factor, w was introduced for each variable, using an ordinary scale from 2 to 1 (2=higher priority, 1=lower priority) based on the experience. Precipitation, temperature, river discharge, and discharge in extraction sites, main canals, water user intakes and irrigation demand in water user fields were established as higher priority variables. Soil texture and depth, crop patterns, storage, and discharge in secondary and tertiary canals are considered as lower priority variables.

A hydrological information availability index, f which is an average of the assigned scores and weighting factors, was determined for the short, medium and long term decisions in irrigation districts according to equation 2.1:

$$f = \frac{\displaystyle\sum_{i,j}^{n}\left(\frac{a_{r_{i,j}} + a_{t_{i,j}} + a_{s_{i,j}}}{m}\right) \cdot w_{i,j}}{\displaystyle\sum_{i,j}^{n} w_{i,j}} \qquad (2.1)$$

Where, f is the hydrological information availability index; a_r is the score assigned for the available period of record; a_t is the assigned score for the available temporal resolution; a_s is the assigned score for the available spatial resolution of hydro-meteorological variable i or discharge information in an established control location j; w is the weighting factor of the hydro-meteorological variable i or discharge information in an established control location j; n is the number of hydro-meteorological variables or control locations evaluated per phase. The value of m depends on the number of variables considered with $m=3$ in the long term planning phase (soil texture and depth, and crop patterns are considered separately) as well as in the medium term planning phase; $m=2$, when evaluating information in the short term operational phase; and $m=1$, when evaluating soil texture and depth, and crop patterns in the long term planning phase.

2.2.3 Selection of irrigation districts

To evaluate the developed index tool, one high-tech irrigation district, the Murrumbidgee district in Australia, and two low-tech irrigation districts, the Coello district in Colombia and the Arenal-Tempisque district (DRAT) in Costa Rica were selected.

The selected irrigation districts are large schemes over 250 km² in size, operating an open canal system with hydraulic structures and gates to regulate the water discharge from the water source (Figure 2.2). Gravity irrigation is predominant, mainly for rice production. In all three irrigation districts, the socio-economic or environmental impacts of making poor water allocation decisions are significant. Operational water management details of each irrigation district are presented in the next section and are summarized in Table 2.4.

Figure 2.2. Typical open canal system with hydraulic structures and gates to regulate the water discharge (DRAT irrigation district).

15

Table 2.4. Selected irrigation districts for hydrological information assessment.

	High-tech		Low-tech
	Murrumbidgee Irrigation district (MI)	Coello irrigation district (Coello)	Arenal-Tempisque irrigation district (DRAT)
Country	Australia	Colombia	Costa Rica
Irrigated area (km²)	1566	250	270
Main crops	Rice, citrus, grapes	Rice, maize, sorghum	Rice, sugar cane, pasture
Main source of water	Murrumbidgee River	Coello River Cucuana River	Arenal reservoir (hydropower operation)
Current type of management	Private agency (Murrumbidgee Irrigation Ltd)	Water users association (USOCOELLO)	Government institution (SENARA)

Arenal-Tempisque irrigation district (DRAT)

In Costa Rica, the Arenal-Tempisque irrigation district (DRAT) is the largest irrigated production area in the country, located in the Guanacaste province with approximately 270 km² of irrigated land (SENARA, 2014). The district is managed by the National Service of Groundwater, Irrigation and Drainage (SENARA), a government institution. The water supply is a low-tech system consisting of two main canals, the South and West Canal, with intakes located upstream of a diversion dam. Short term decisions on opening or closing the gates in these intakes are taken based on the discharge information manually measured at the diversion dam. In water abundant conditions, a minimum discharge is set for supplying the South Canal (16 m³/s), due to aquaculture activities. This flow is returned to the canal system in its entirety, and can be used for satisfying the irrigation demand. In water scarce conditions, 30% of the available water is allocated to the South Canal, and 70% of the available water is allocated to the West Canal. Currently, the South Canal is being extended to satisfy irrigation demand in an additional production area in the district (SENARA, 2014), and there are long term plans to build a reservoir that can alleviate water shortages.

The water availability in the district is constrained by the releases from an upstream reservoir, which is operated by the Costa Rican Institute of Electricity (ICE). Normally, there is sufficient water available to satisfy irrigation demand. However, when storage is prioritized for hydroelectric energy production, a reduced release discharge is made available for DRAT. Releases from the reservoir can instantly change, due to water storage priorities in the upstream reservoir, leading to water scarcity at the diversion dam. Currently, the monthly irrigation demand is determined using climatological precipitation data, crop evapotranspiration estimates and the estimated total system efficiency. Irrigation demand estimates are updated per semester according to the change in crop patterns. Rice is often grown in rotation with sugarcane and pasture. The ability of the soil to pond water without excessive contribution to the groundwater or environmental effects to surrounding lands are factors in approving areas suitable for rice production (SENARA, 2014).

Coello irrigation district (Coello)

The Coello irrigation district is located in the upper Magdalena macro-basin in Colombia, in the Tolima Department, a region vulnerable to droughts and climate variability including the ENSO climate phenomena (IDEAM, 2015).

The district is managed by a water user association (USOCOELLO) and the river diversion system serves an irrigated area of approximately 250 km², comprising mainly of irrigated rice (Urrutia-Cobo, 2006). The water availability for the irrigation district depends on the runoff of two rivers (Cucuana and Coello Rivers), supplying two main canals. The water supply in these canals is a low-tech system, with the opening and closing of control gates as well as measuring of flows in the canals carried out manually.

Short and medium term water allocation decisions are taken based on the measured discharge in the main canal intakes, with the average irrigation module estimated at 0.2 $m^3/s/km^2$ for rice production. Drought warnings issued by the national meteorological institute (IDEAM) may be considered. Irrigation requests are approved to the extent that the predicted water availability meets the demand. Currently, irrigation demand is based on the irrigation module estimate. Water is allocated to farmers on the basis of area, and crop type (e.g. rice, maize, sorghum or cotton), but it may be constrained in water scarce situations. When the water availability is insufficient for planting rice over the entire system, the association introduces a rice rotation and zoning plan to enable all farmers to plant rice at least once per year (Urrutia-Cobo, 2006; Vermillion and Garcés-Restrepo, 1996).

Murrumbidgee irrigation district (MI)

The Murrumbidgee irrigation district is located in the Murray-Darling basin in Australia. In this region, climate variability is one of the greatest sources of risk to agriculture (Richter, 2014). Currently, there is a high-tech decision support system in place with real-time discharge measurements in canals and an online water ordering system to avoid agricultural production loss due to water shortage. The Murrumbidgee Irrigation Ltd is the company responsible for supplying water to satisfy irrigation demand in the district. It manages the water supply service through the use of a system of modelling tools and remote monitoring, and the control of critical infrastructure to allow coordination in operating the supply system.

The company provides water entitlement contracts and water delivery contracts to the users of an irrigated area close to 1566 km^2 (MI, 2013a). The company determines the level of service that can be provided for the medium term planning phase according to water availability, the priority of water orders among pricing groups, the availability of the company's works for the delivery of water allocation on different dates throughout the year and access to the company's drainage works. For the short term operational phase, the time of day for starting and stopping of delivery of allocated water are set in accordance with water orders. These water orders are established by the users in an online order system with specific information requirements, including water demand and use; the requested start and finish time, and date for delivery. The online system allows monthly updated information to be obtained on the amount of water delivered to the user according to ground measurements (MI, 2013a).

Rice is the most dominant water user, alongside other agricultural products like grapes and citrus (Khan et al., 2006). Rice is often grown in rotation with leguminous pastures and dryland crops, due to limited water availability for irrigation. The company has developed special rules for authorizing irrigated rice production (MI, 2013b), considering a soil assessment of the field.

The volume of water available for the irrigation district is determined by considering the water levels in two reservoirs (Blowering and Burrinjuck) in the upper Murrumbidgee basin. Based on a forecast system, decisions on releasing water from the reservoirs are taken, and the amount of water available for irrigation is set by the New South Wales (NSW) Government. The available water is diverted from the main offtake (Berembed Weir) to the Bundidgerry storage and regulator, the start of the irrigation canal system (MI, 2015b).

2.2.4 Determining the rate of occurrence of sub-optimal water allocation decisions

The rate of occurrence of sub-optimal water allocation decisions was defined as the percentage of time in a year where water was available to satisfy irrigation demand, but not supplied accordingly, thus generating water shortage in the main supply system. This implies that the water shortage occurred due to the decision to not supply the full demand, even though the water was available.

In contrast, an optimal decision is made if the water supply is curtailed when there are water scarce conditions (optimal decisions Y in Table 2.5), as well as in the case where the supply is set to meet the demand under non-water scarce conditions (optimal decisions N in Table 2.5). The condition of taking the decision to meet the demand under conditions of water scarcity cannot exist due to lack of water. Contingency tables were generated to evaluate the decisions made on water allocation in water scarce and water abundant conditions (Table 2.5). Available ground data and information of discharge at the extraction sites (Q), water supply in the main canals (S), and irrigation demand estimates (D) were used for this purpose (Table 2.5).

Table 2.5.   Contingency table: A decision framework for water allocation at main canal level in irrigation districts.

	Action 1: Generating water shortage due to decision $(S-D)<0$	Action 2: Generating **no** water shortage due to decision $(S-D)\geq0$
State 1: Water scarcity $(Q-D)<0$	*Optimal decisions Y* *(Scarcity/Shortage)*	*Not applicable*
State 2: No water scarcity $(Q-D)\geq0$	*Sub-optimal decisions* *(No scarcity/Shortage)*	*Optimal decisions N* *(No scarcity/No shortage)*

During operation of the Arenal-Tempisque irrigation district (DRAT), decisions concerning opening and closing the main gates of the independent canal systems (South Canal and West Canal) are taken at one extraction site (Figure 2.3). Water shortage was defined as a water deficit in either one of the two main canals, where the measured supply is lower than the estimated irrigation demand in the command area of that canal. Water scarcity in the DRAT district was defined as the water deficit at the extraction site, where the availability of water is lower than the estimated irrigation demand in the command area of both canal systems.

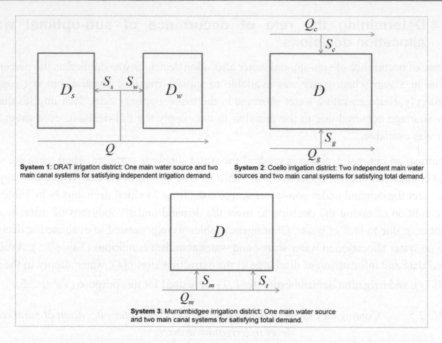

Figure 2.3. System configuration at main canal level: Discharge availability Q, supply S and demand D in irrigation districts DRAT, Coello and Murrumbidgee.

Observed daily water availability at the extraction site (Q), daily supply in the two main canals (S_s and S_w), and monthly irrigation demand for each main canal sector (D_s and D_w) are available from SENARA and were used to establish the contingency table for five years (2008-2012).

Currently, monthly irrigation demand is established using precipitation and evapotranspiration estimates, as well as information on crop patterns and estimated efficiency (SENARA, 2014). However, irrigation demand can vary due to dry or wet conditions. The assumption was made, that the irrigation demand could be 25% higher or lower, depending on the precipitation deficit or surplus respectively. Precipitation anomalies were obtained from the meteorological reports of the *Instituto Meteorológico Nacional* (IMN), Costa Rica (IMN, 2014).

In the Coello irrigation district, decisions on the opening and closing of the main gates are taken at the two extraction sites in the Cucuana and Coello Rivers. Even though these extraction sites are independent, the available water is supplied to the same command area (Figure 3). Therefore, in the Coello district, water shortage was defined as the water deficit in the main canal, when the total supply from the two extraction sites is lower than the estimated irrigation demand in the command area. Water scarcity was defined as the deficit of available water in the rivers at the combined extraction sites as compared to the total irrigation demand.

Water supply measurements in the main canals (S_c and S_g) and estimates of irrigation demand (D) were obtained based on data and information provided by USOCOELLO. Water availability in the rivers at the two extraction sites (Q_c and Q_g) were estimated based on discharge data provided by the *Instituto de Hidrología, Meteorología y Estudios Ambientales* (IDEAM), Colombia. Contingency tables were established for three years (2010-2012). As in the DRAT irrigation district, the irrigation demand was adjusted as a function of reported precipitation anomalies. These were obtained from the meteorological reports of IDEAM (IDEAM, 2015).

In the Murrumbidgee irrigation district, decisions on the opening and closing of the main gates are taken at the two extraction sites in the Murrumbidgee River. The available water is supplied to one command area (Figure 3), thus water shortage and water scarcity were defined similarly as for the Coello irrigation district.

Observed daily water availability (Q_m) and daily supply in the two main canals (S_m and S_t) were obtained based on data provided by the New South Wales (NSW) Government, Department of Primary Industries Office of Water. Daily irrigation demand was estimated based on the water announcements made by the NSW Government between the years 2004 and 2008. Contingency tables were established for those five years.

2.3 RESULTS

2.3.1 Information availability of hydro-meteorological variables for the long term planning phase

An average score per hydro-meteorological variable (precipitation, temperature, runoff, storage, soil texture and depth, and crop patterns) was determined according to the temporal and spatial availability of data for the long term planning phase (Figure 2.4). The irrigation districts of Coello and DRAT (Colombia and Costa Rica) have similar results. For both of these irrigation districts, the average score obtained for the availability of precipitation data is 3.7 (Figure 2.4). This score accounts for the availability of precipitation data being fair to good, with similar temporal, spatial and period of record availability for the two districts. For temperature and runoff variables, the availability of information is fair, with an average score close to 3 (Figure 2.4). Reasonable spatial information was found for cropping patterns (scores between 4 and 5), but information on soil texture and depth is limited (Figure 4). In both districts there is no significant storage that can be used to satisfy irrigation demand, thus a score of zero was assigned to the storage variable. In the DRAT district, there is a storage reservoir in the river upstream of the district, but its operation does not prioritize irrigation.

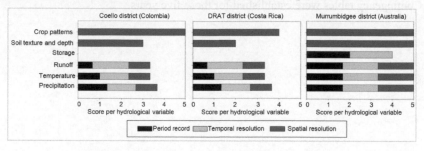

Figure 2.4. Scores per hydro-meteorological variable for long term planning phase in the selected irrigation districts.

In the Murrumbidgee irrigation district, one key storage reservoir is located at the beginning of the main canal. Good (4) temporal resolution and period of record were found for storage information. For all other hydro-meteorological variables the availability of information is considered excellent (Table 2.3).

2.3.2 Information availability of discharge in control locations for medium term planning and short term operational phase

An average score for discharge information in water extraction sites, canals and water user's intakes, and irrigation demand in fields, was determined according to the temporal and spatial evaluation (Figure 2.5, Figure 2.6). The findings show that availability of information is similar for the Coello and DRAT systems, except that river discharge information is available for the DRAT. However, the period of record of the river discharge in the DRAT district is limited (score of 2), which reduces the average score for the medium term to 3.7, although the spatial resolution is good (score 4) and the temporal resolution is excellent (Figure 2.5). The evaluation of the same information for short term operational phase shows good spatial and temporal resolution, which results in an average score of 4.0 (Figure 2.6).

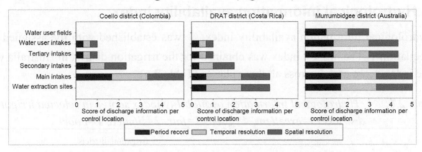

Figure 2.5. Scores for discharge information per control location, for the medium term planning phase in the selected irrigation districts.

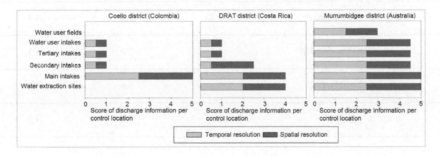

Figure 2.6. Score for discharge information per control location, for the short term operational phase in the selected irrigation districts.

Both irrigation districts have reasonable temporal and spatial discharge information in the main canal intakes, but Coello has a longer period of record, with over 30 years of monthly data available. Therefore, the average information availability score for the medium term planning phase in Coello is higher in the main canal intakes (Figure 2.5). Discharge information in the secondary canal intakes is also better, due to a longer period of record. Both irrigation districts show poor discharge information availability in the tertiary canals and at the water users intakes. In addition, the spatial and temporal resolution of hydro-meteorological variables is not adequate to determine reliable irrigation demand at field level. Hence, information availability

23

for estimating the irrigation demand is considered limited for both districts. The information availability for the short term operational phase is considered poor, based on the results obtained in the secondary and tertiary canals and at the water user's intakes (Figure 2.6). Better information availability is found in the main canal intakes (scores 4 to 5).

For the Murrumbidgee irrigation district, the information availability scores range between fair (3) to excellent (5). For the long term planning phase, all hydro-meteorological variables have excellent information availability. For the medium and short term phase, the availability of information is the highest at the main intakes and water extraction sites (Figure 2.5 and Figure 2.6), followed by the information at the secondary, tertiary and water user intakes with an average score of 4.5. The score for discharge information from user's fields is fair (score of 3).

2.3.3 Hydrological information availability index

The hydrological information availability index, f was established with the assigned scores using equation 1. The highest index was obtained for the irrigation district in Australia with an average index score of 4.5 across all three phases (Table 2.6).

Table 2.6.   *Hydrological information availability index scores in selected irrigation districts (5=excellent, 4=good, 3=fair, 2=limited, 1=poor).*

Hydrological information availability index, f	Irrigation districts		
	Coello (Colombia)	DRAT (Costa Rica)	Murrumbidgee (Australia)
f_{so} short term operation	1.4	2.2	4.4
f_{mp} medium term planning	1.5	2.0	4.3
f_{lp} long term planning	3.3	3.6	4.9
f average	2.1	2.6	4.5

An average information index of 2.1 and 2.6 was established for the Coello and DRAT systems, respectively, implying a poor (1) to fair (3) information availability for all phases (Table 2.6). However, differences between the two low-tech irrigation districts are interesting. The Coello district has less hydrological information for the short-medium phase, but has similar information availability for the long term planning phase. The index scores for the short and for the medium term for Coello are 1.4 and 1.5, respectively, and 2.2 and 2.0, respectively for DRAT (Table 2.6). For these phases, measurements in secondary, tertiary and field intakes have to be improved both spatially and temporally. For the long term planning phase, the index results are relatively better (3.3 for Coello and 3.6 for DRAT), due to the availability of key variables, such as precipitation and temperature. Even though the score for the availability of precipitation and temperature data is fair to good, only one climate station is currently being used in the DRAT district to estimate irrigation demand. Irrigation demand estimates at the sub-district or field level can therefore be misleading, affecting both planning and operational water allocation decisions. Efforts should be made to improve the climate station density and the use of this data for the estimation of irrigation demand.

2.3.4 The rate of occurrence of sub-optimal water allocation decisions

After establishing the information index for the selected irrigation districts the rate of occurrence of sub-optimal and optimal water allocation decisions was evaluated as explained in section 2.4.

Similar results were found for the sub-optimal and optimal water allocation decisions taken in Coello and DRAT irrigation districts. In the DRAT irrigation district, results show that the rate of occurrence of sub-optimal water allocation decisions was between 10% and 24% of the time per year (Figure 7). The rate of occurrence of optimal water allocation decisions (Y), under true water scarcity conditions, was between 2% and 19% of the time and the optimal water allocation decisions (N), under no water scarcity conditions between 63% and 85% of the time. In the Coello irrigation district, the rate of occurrence of sub-optimal water allocation decisions was between 16% and 36% of the time per year with corresponding rates of optimal water allocation decisions being taken as shown in Figure 2.7.

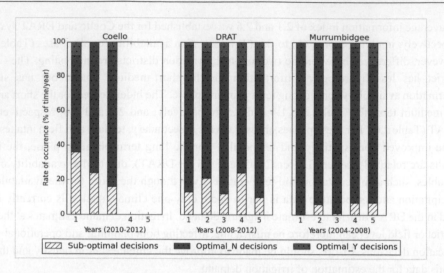

Figure 2.7. The rate of occurrence of sub-optimal and optimal water allocation decisions in irrigation districts: Coello (2010-2012), DRAT (2008-2012) and Murrumbidgee (2004-2008).

In the Murrumbidgee irrigation district, in the year 2004, a rate of 0% of occurrence of sub-optimal water allocation decisions was obtained (Figure 7). In the year 2007 and 2008, the rate of occurrence of optimal water allocation decisions Y, under true water scarcity conditions was 3% and 13% of the time. For those years, sub-optimal water allocation decisions were taken 7% of the time, a lower rate compared to Coello and DRAT irrigation districts.

These results, imply that there is more room to improve water allocation decisions in DRAT and Coello irrigation districts given the current hydrological information availability.

2.4 DISCUSSION

2.4.1 Linking the hydrological information index and sub-optimal water allocation decisions

For the selected irrigation districts, a link was found between hydrological information index results and the rate at which sub-optimal water allocation decisions are being taken. The proposed relationship between the information index scores, the rate of occurrence of sub-optimal water allocation decisions, and the marginal value of additional information is presented in Figure 2.8 and Figure 2.9.

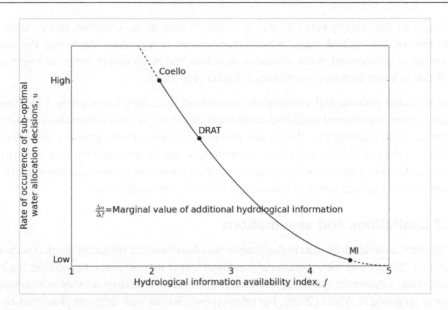

Figure 2.8. The rate of occurrence of sub-optimal water allocation decisions against the score of the hydrological information availability index in the selected irrigation districts (Coello, DRAT, and MI).

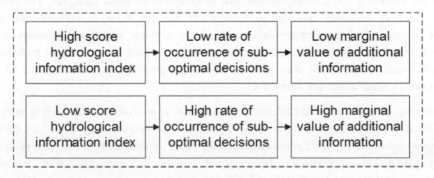

Figure 2.9. The relationship between the score of the hydrological information index, the rate of occurrence of sub-optimal water allocation decisions, and the marginal value of additional hydrological information.

When the information index score is high, such as a value of 4.5 in the Murrumbidgee irrigation district (abbreviated to MI), the rate of occurrence of sub-optimal water allocation decisions is found to be low. When the information index score is low, (e.g. 2.1 or 2.6 in the Coello and DRAT irrigation districts, respectively), the rate of occurrence of sub-optimal water allocation decisions is found to be high.

This suggests that, having better hydrological information in the irrigation district leads to a lower rate of sub-optimal water allocation decisions being taken. Reducing the rate of occurrence of sub-optimal water allocation decisions makes economic sense, as lower water deficit due to better decisions contributes to higher crop yields.

If the available hydrological information is excellent (e.g. high index score 4.5), then the marginal value of additional hydrological information is low, as more information cannot have a major effect in reducing the, already low rate of occurrence of sub-optimal water allocation decisions. If the available hydrological information is poor or intermediate, then the marginal value of additional hydrological information is high, as more information can have an effect in reducing the rate of occurrence of sub-optimal water allocation decisions.

2.4.2 Limitations and assumptions

The information index presented in this chapter was developed for irrigation districts with areas larger than 250 km² and with operational periods of longer than 30 years. In addition, the index tool considers a minimum spatial representativeness for river discharge stations in mountainous basins as proposed in WMO (2008). For other types of basins with different characteristic, the ranges used in scoring can be adjusted.

As it is applied here, the limits for precipitation and temperature data have also been taken as the same, assuming that climate stations will include both of these variables. The approach could, however, be easily extended to differentiate between these, as normally the greater spatial correlation of temperature data would imply a lower spatial representativeness of stations would be sufficient. Other variables may also be taken into account, including for example water quality indicators such as salinity, depending on the relevance of these parameters in water allocation decisions.

Additionally, sub-optimal water allocation decisions were assessed as those where water was not allocated (extracted) despite it being available. Maintenance (e.g. repair of hydraulic structures) or other constraints could have been the cause of the closing of the main gates of the supply system. This information is, however, not available in the irrigation districts assessed, and the assumption was taken that the sub-optimal decisions do occurr as a result of the scarcity of hydrological information in determining water availability and irrigation demand. Obviously, maintenance has to be done during the operation of the irrigation districts, but the additional hydrological information would also serve to better plan maintenance if it were available, avoiding the occurrence of shortfalls of meeting the demand.

2.4.3 Usefulness for irrigation districts and further work

The proposed index tool provides irrigation managers a measure as to the resolution and period of record at which information should ideally be available. Ensuring information of sufficient quality is available will promote improved decisions for water allocation in agricultural production.

Better water allocation decisions in irrigation districts would require additional hydrological information. Additional hydrological information can be obtained by expanding networks of gauging and climatological stations, but may also be based on remote sensing, hydrological and meteorological models (Karimi and Bastiaanssen, 2015; Peña-Arancibia et al., 2014, 2016). Peña-Arancibia et al. (2016) suggest that surface water use estimates can be improved using these datasets when compared to traditional crop modelling. Combining remote sensing and hydrological models may help improve the index scores and enhance water allocation decisions. However, it is crucial to establish an adequate spatial and temporal resolution according to the operational and planning phases in the district, and then explore how that data can be best provided. Further research can also help establish the marginal value of this improved information in a quantitative manner by estimating the change in crop yield in irrigation districts due to reducing sub-optimal decisions.

2.5 CONCLUSION

Large irrigation districts are the largest water consumers in the planet. A tool to assess the availability of hydrological information and the rate of occurrence of sub-optimal water allocation decisions was developed for these water supply systems.

Sub-optimal water allocation decisions are considered as those where the supply to meet the demand was curtailed, despite sufficient water being available. It was found that in the selected irrigation districts in Colombia, Costa Rica and Australia (Coello, DRAT, and Murrumbidgee irrigation district), sub-optimal water allocation decisions do occur.

An index tool was applied to assess the availability of hydrological information for different decision phases; long term planning, medium term planning and short term operational phases. The quality of that information was evaluated by scoring the period of record; and the temporal and spatial resolution of the available data (e.g. precipitation, temperature, and discharge). Of the three irrigation districts considered, the one in Australia (Murrumbidgee irrigation district) was found to have excellent hydrological information availability, while the selected irrigation districts in Costa Rica (DRAT irrigation district) and Colombia (Coello irrigation district) have fair to poor availability of information.

In the DRAT and the Coello irrigation districts, the rate of occurrence of sub-optimal allocation decisions was found to be higher (10% to 36% of the time in one year) compared to the Murrumbidgee irrigation district (0% to 7% of the time in one year) as a consequence of lower availability of hydrological information. The rate of occurrence of sub-optimal water allocation decisions increases as the score of the hydrological information index decreases. The index score can potentially be used as a proxy of the rate of occurrence of sub-optimal water allocation decisions. This tool provides irrigation managers a measure as to the resolution and period of record at which information should ideally be available. Ensuring information of sufficient quality is available will promote improved decisions for water allocation in agricultural production. Though its application would need to be validated in other irrigation districts.

The results imply that the use of additional hydrological information is beneficial in reducing the rate of occurrence of sub-optimal water allocation decisions, which ultimately contribute to higher crop yields. The marginal value of this additional information can be determined by evaluating the change in the rate of occurrence of sub-optimal water allocation decisions and the change in available hydrological information.

The challenge is to incorporate additional hydrological information through additional ground stations, remote sensing, hydrological models and global models. The contribution of such additional data is currently being addressed as part of this ongoing research.

3

THE BENEFIT OF USING AN ENSEMBLE OF GLOBAL HYDROLOGICAL MODELS IN SURFACE WATER AVAILABILITY ESTIMATES

Based on: *Kaune A., López-López P., Gevaert A., Veldkamp T., Werner M., de Fraiture C.: The benefit of using an ensemble of global hydrological models in surface water availability for irrigation area planning. Water Resour. Manag. Rev., 2018. (In review).*

Abstract

Hydrological data and information on the availability of water are essential to support water allocation decisions in irrigated agriculture, especially under increasingly water scarce conditions. However, in many agricultural regions around the world hydrological information is scarce, leading to sub-optimal water allocation decisions and potential agricultural production loss. In this study we assess the influence of hydrological variability on water availability, and evaluate the benefit of using an ensemble of global hydrological models for designing the irrigation area, where this is determined based on agreed operational targets of water supply reliability. Surface water availability estimates are established using an ensemble of global hydrological models. The risky outcomes in terms of annual agricultural production due to the actual frequency of occurrence of water scarcity were generated for a reference and alternative irrigation areas, resulting in a Relative Utility Value (*RUV*) that expresses the utility of the information used. Results show that using an ensemble of global hydrological models provides more robust estimates of the planned area compared to when using any of the single global models that constitutes the ensemble. A comparison of the information content in the ensemble shows that an ensemble with a period of record of fifteen years has an information content equivalent to a single model of thirty years. The results provide insight into how a hydro-economic framework can be applied in irrigation area planning, given uncertain surface water availability estimates.

3.1 INTRODUCTION

The debate on water scarcity at the global level centres on population growth and per capita income increases (FAO, 2013; Rijsberman, 2006; Wada et al., 2016), exacerbated by the uncertain availability of water due to climate variability and climate change (Eslamian, 2014; Gosling et al., 2017; Lutz et al., 2014). Adequate prediction tools for water resources assessment are key to provide a better understanding of present and future water scarcity, and its impacts on human livelihoods, the environment and agricultural development; especially in regions where hydrological data is scarce (Bloschl et al., 2014; Masafu et al., 2016; Tegegne et al., 2017). In these regions the potential of using additional hydrological information from global hydrological models is high as these can help enhance water resource availability assessments, and decrease the number of sub-optimal water allocation decisions (Kaune et al., 2017).

Several global hydrological models have been developed (Bierkens, 2015; Kauffeldt et al., 2016; Sood and Smakhtin, 2015) and have been widely applied to climate change and water scarcity impact assessment, and improved river discharge estimations (van Beek et al., 2011; Gosling et al., 2017; Hanasaki et al., 2013; Lopez Lopez, 2018; Veldkamp et al., 2015a; Zhao et al., 2017). Global hydrological models such as LISFLOOD (Knijff et al., 2010), WATERGAP3 (Döll et al., 2009; Flörke et al., 2013), PCR-GLOBWB (van Beek et al., 2011; Wada et al., 2014), SURFEX-TRIP (Decharme et al., 2010, 2013) and HTESSEL (Balsamo et al., 2009) provide surface water availability estimates for large regions and basins. These models are typically forced using meteorological re-analysis data such as ERA-Interim (Dee et al., 2011), or the more recently developed MSWEP (Beck et al., 2017b), that integrates in-situ meteorological data, global earth observations and the ERA-Interim reanalysis. Obviously locally calibrated hydrological models may be more suitable at river basin scale (López López et al., 2016; Zhang et al., 2016), but these depend on the availability of observed data from in-situ stations with sufficient quality and period of record, which in many areas of the world may be an issue. Although large-scale hydrological models provide consistent data with a long period of record, there are large differences between outputs of the different models (Gudmundsson et al., 2012; van Huijgevoort et al., 2013; Stahl et al., 2012). Studies have found that using a model ensemble can be used to increase the accuracy of predicting hydrological variables (Gudmundsson et al., 2011; van Huijgevoort et al., 2013), resulting in more reliable surface water availability estimates. These studies explore the use of ensembles of global models for water resources assessment, the focus is on the gain in selected statistical performance metrics. Other studies have evaluated the gain or benefit of the use of ensemble information from an economic aspect. Verkade and Werner (2011) estimate the economic benefit of using forecasts for flood warning, while (Quiroga et al., 2011) evaluate the economic benefit of using drought information for water allocation decisions; both applying the theory of Relative Economic Value (Murphy, 1985). Other studies that propose hydro-economic frameworks to support decisions in water resources systems include Alfonso and Price (2012), Bouma et al. (2009) and Macauley (2006), who apply the expected utility theory (Neumann and

Morgenstern, 1966) for estimating the value of information through a Bayesian setting (Hirshleifer and Riley, 1979).

While hydro-economic frameworks have been applied to estimate the benefit of using hydrological information for several applications, a gap exists in supporting decisions in agriculture, especially in large irrigation districts (Kaune et al. 2007). Irrigation districts are hydro-productive schemes designed to provide a water service for agricultural production. Operators in the scheme decide when and where to supply water to secure crop yields in the irrigated area. If the available water is not sufficient to satisfy crop water requirements, then the operational system would have failed in providing the necessary service, leading to potential agricultural losses due to water scarcity. Planning the adequate irrigation area based on solid information about the water availability is crucial since the initial design phase of the irrigation district. Data on river discharge, groundwater, soil types, and desired cropping patterns is required for the water balance assessment. Investors would like to receive the best irrigation area advice If water resources are overestimated the higher frequencies of water scarcity may occur than planned for. Alternatively if the water resources are underestimated, then the irrigation area may be too small, leading to an opportunity cost. We apply a risk based approach to designing irrigation areas using a hydro-economic Expected Annual Utility framework based on the economic principles of Neumann and Morgenstern (1966), assuming risky outcomes that consider the occurrence and non-occurrence of water scarcity.

We hypothesise that the use of an ensemble of global hydrological models provides more reliable estimates of surface water availability for irrigation area planning compared to individual models. This is evaluated through the hydro-economic framework in a selected basin that determines the value of using the model ensemble in determining the areas that can be irrigated as a function of the estimated water availability.

3.2 MEHTODS

The Relative Utility Value *RUV* used in this study is defined as the risky annual crop production given monthly probabilities of (non-)water scarcity compared between the reference and the planned irrigation areas. This value includes the irrigation areas for river discharge simulations derived using different global hydrological models, the monthly probability of water scarcity using these areas, and the potential yield reduction due to water deficit for rice. The framework was tested in a basin in Southeast Australia.

3.2.1 Murrumbidgee River basin

We apply our analysis to the Murrumbidgee River basin in Australia (Figure 3.1). This basin was selected because it has a dense network of in-situ stations, which can be used to compare model results.

33

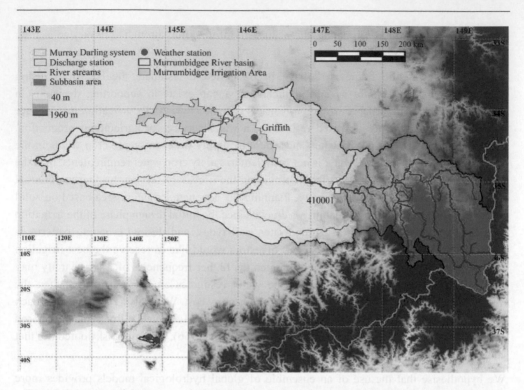

Figure 3.1. Map of the Murrumbidgee River basin in the Murray Darling System, Australia. Location of the Murrumbidgee Irrigation Area (MIA), weather and discharge stations.

The Murrumbidgee basin is a highly regulated system located in southeast Australia (State of New South Wales) and supplies water for agriculture production, safe drinking water, industry and the environment. The basin is part of the Murray-Darling system, which was heavily impacted by the Millennium drought between 2001 to 2009 (van Dijk et al., 2013b). In 2012, The Murray-Darling Basin Authority released a new Basin Plan for the integrated management of water resources to prevent future major impacts due to extreme climatological events. Water allocation decisions are made by the New South Wales government throughout the season, considering water availability primarily based on water levels in two major reservoirs (Burrinjuck and Blowering) in the upper part of the basin with a combined volume capacity of $2.654x10^9$ m³. The river flows from east to west from approximately 1960 m to 40 m elevation, with a high hydroclimate variability with a mean annual discharge of 125 m³/s (Green, 2011). At the Wagga Wagga weather station (middle of the basin) the precipitation varies between 35 mm/month and 55 mm/month. Months between November and February are the driest (summer season). In the mountainous higher part of the basin the precipitation is two times higher than the precipitation in the middle of the basin (Burrell, 2017 and Green 2011). The hottest months in the basin are January and February when temperature vary from 33^0C in the west and 16^0C at higher altitudes in the east. Winter temperature vary from 3^0C to 5^0C in the west to 0^0C to - 2^0C in the east (MWRP, 2017). The evaporation is higher at lower elevation with 1800mm/year in the west and 1000 mm/year in the east. At Wagga Wagga weather station, in the winter period during July evaporation may be as low as 1 mm/day, while in summer during January it reaches 9 mm/day (Green, 2011).

Major irrigation districts have been developed in the middle of the basin (west) with a total command area estimated at 4182 km² (reference green). The largest district is the Murrumbidgee Irrigation Area (MIA) with approximately 3624 km² (Green, 2011). Irrigated agriculture in the basin provide 25% of the state's fruit and vegetable production and half of Australia's rice production. Rice is the dominant water user (MI, 2015a), with a growth period from November to February. In this research, we focus only on rice production with corresponding four months of total growth length and crop yield factors for each growth stage representing the sensitivity due to water deficit.

The Murrumbidgee basin is a highly regulated basin with two large reservoirs and a well-established water allocation decision process. It is a complex system which requires the calibration of several model parameters for accurate river discharge simulations. However, in this chapter, we want to assess the information content of the ensemble of the global hydrological models. The observed river discharge in the Murrumbidgee basin has been naturalized to avoid the influence of the reservoir operation. Hence, allowing the use of the global hydrological models in the non-reservoir operation model.

3.2.2 Hydro-meteorological data

Precipitation data was obtained from the Multi-Source Weighted-Ensemble Precipitation, MSWEP (Beck et al., 2017b). Air temperature data were based on the WATCH Forcing Data methodology applied to ERA-Interim reanalysis, WFDEI (Weedon et al., 2014) corrected with the CRU data (Harris et al., 2014). Precipitation and air temperature datasets were available for 1980-2009 period at 0.25° x 0.25° spatial resolution and daily temporal scale.

Monthly river discharge data was obtained from the Australian Bureau of Meterology (BoM) at the Wagga Wagga station (410001) for the 1980-2009 period, which is upstream of the intakes to the main irrigation areas in the basin (Figure 1). As the basin is highly regulated, the river discharge at Wagga Wagga station was naturalised using the upstream hydrological reference stations (Turner, 2012; Zhang et al., 2014), which are in unregulated headwater catchments. The stations used to naturalise the streamflow are: Cotter River at Gingera (410730), Adelong Creek at Batlow Road (410061), Goobarragandra River at Lacmalac (410057), Gudgenby River at Tennent (410731), Molonglo River at Burbong (410705) and Murrumbidgee River below Lobbs Hole Creek (410761). The naturalised flow was obtained with the monthly discharge of the hydrological reference stations adjusted with the quotient of the mean discharge at the Wagga Wagga station and the mean discharge of the hydrological reference stations. The naturalised flow is used as the reference surface water availability for irrigation, though an environmental flow of 25% of the river discharge was considered.

3.2.3 Global hydrological models

Eight different global hydrological and land surface models were used: HTESSEL (Balsamo et al., 2009), SURFEX-TRIP (Decharme et al., 2010, 2013), PCR-GLOBWB (van Beek et al., 2011; Wada et al., 2014), WATERGAP3 (Döll et al., 2009; Flörke et al., 2013), LISFLOOD (Knijff et al., 2010), ORCHIDEE (Krinner et al., 2005), JULES (Best et al., 2011), and W3RA (van Dijk et al., 2013a). For each model the total runoff was obtained on a 0.25° x 0.25° spatial resolution, and all models were forced with a consistent meteorological forcing dataset (see section 2.2). Monthly runoff simulations for each model were obtained over the 30-year study period from 1980 through 2009 for the basin to the station at Wagga Wagga and compared with the naturalized observations at monthly time scale for the 1980-2009 period. Several performance metrics were established to evaluate the hydrological performance of the models, including Kling-Gupta efficiency (KGE, Gupta et al., 2009); Pearson's correlation coefficient (r), Root Mean Square Error (RMSE) and percent bias.

For each global hydrological model the monthly surface water availability for irrigation was derived from the simulated runoff for the 1980-2009 period. The model ensemble was made by combining the monthly surface water availability from all eight models. This means that the ensemble contains eight surface water availability values for each month to explicitly include the model uncertainty. To explore the influence of the period of record in determining the surface water availability, the samples with a different number of years are extracted from the thirty year dataset. This resulted in independent groups of data for 5 years (1980-1984, 1985-

1989, 1990-1994, 1995-1999, 2000-2004, 2005-2009); 10 years (1980-1989, 1990-1999, 2000-2009); 15 years (1980-1994, 1995-2009) and the full 30 years (1980-2009) (Figure 3.2).

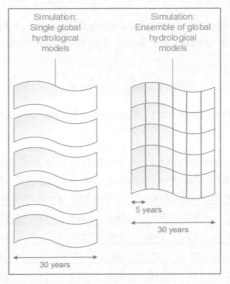

Figure 3.2. Obtaining simulations of single global hydrological models for 30 years and an ensemble of global hydrological models for 5, 10, 15 and 30 years.

3.2.4 Determining the irrigation areas

The irrigation area is determined based on an operational target of water supply reliability (75%) using a bootstrap resample of the multi-annual monthly river discharge datasets to constrain the statistical uncertainty. The reference irrigation area is established using the observed (naturalised) surface water availability for 30 years (1980-2009), while the planned irrigation areas sare determined using each simulation of surface water availability for the selected periods (mentioned in section 3.2.3.

Water supply reliability, or probability of non-occurrence of water scarcity $pr\{NWS\}$ is defined as the relative frequency p of the multi-annual monthly surface water availability Q_a being able to satisfy the irrigation demand D (Equation 3.1a and 3.1b).

$$pr\{NWS\} = 75\% \tag{3.1a}$$

$$pr\{NWS\} \leq p\{(Q_a - D) \geq 0\}_{month} \tag{3.1b}$$

The irrigation demand D (equation 3.2) varies depending on the size of the irrigation area A, which is the variable to be obtained, and a fixed monthly irrigation demand rate r. The monthly irrigation demand rate (Table 3.1) was established using average precipitation and potential evapotranspiration data from a local weather station (MI, 2015a), and a total irrigation efficiency of 77% (Khan et al., 2006).

37

$$D = A \cdot r \tag{3.2}$$

For each irrigation area size that is obtained using the simulated data, the real probability of water scarcity $pr\{WS\}$ is determined using the observed surface water availability $\hat{Q}_a$ (equation 3.3).

$$pr\{WS\} = p\left\{\left(\hat{Q}_a - D\right) < 0\right\}_{month} \tag{3.3}$$

These probabilities are then used for evaluating the benefit of using the ensemble of global hydrological models.

Table 3.1. Irrigation demand rate, r determined with the precipitation, P and potential evapotranspiration, ETP data from weather station Griffith (CSIRO) and using a total irrigation efficiency of 77%. Units are in mm/month.

	Jul	Aug	Sep	Oct	Nov	Dec	Jan	Feb	Mar	Apr	May	Jun
ETP	42	66	100	153	194	238	250	194	162	91	53	36
P	34	35	35	40	32	30	33	32	36	31	37	35
r	10.4	40.3	84.4	147	210	270	282	210	164	77.9	20.8	1.30

3.2.5 Evaluating the benefit of using the ensemble of global hydrological models

The benefit of using the ensemble of global hydrological models was evaluated with the Relative Utility Value (RUV) an index composed of Expected Annual Utility (U) estimates. According to economic theory (Neumann and Morgenstern, 1966), the Expected Annual Utility U is defined as the sum of Risky Outcomes R for the occurrence and the non-occurrence of an event (equation 3.4).

$$U = R_{WS} + R_{NWS} \tag{3.4}$$

In this study, the Risky Outcomes R_{WS} and R_{NWS} are determined as the annual agricultural production due to occurrence and non-occurrence of water scarcity using the real probabilities of water scarcity $pr\{WS\}$, the potential annual agricultural production that can be achieved under non water scarcity conditions P_{NWS}, and the annual agricultural production loss L_{WS} due to water deficit happening in a given month (Equation 3.5 and Equation 3.6).

$$R_{WS} = pr\{WS\} \cdot \left(P_{NWS} - L_{WS}\right) \tag{3.5}$$

$$R_{NWS} = \left(1 - pr\{WS\}\right) \cdot P_{NWS} \tag{3.6}$$

The annual agricultural production loss is determined with Equation 3.7, assuming water deficit happens independently in each month. A simplified approach is used where each month corresponds to one growth stage of the crop. The expected crop yield without water deficit is established according to local agricultural information, while the final crop yield is obtained using yield reduction values based on known yield response factors per growth stage (FAO, 2012) and the actual occurrence of water deficits. For larger or smaller irrigation areas than the reference area, the evapotranspiration reduction changes proportionally as the available water is uniformly distributed in the new irrigation area.

$$L_{WS} = c \cdot A_{planned} \cdot \sum_{month} (y_o - y_f)$$ (3.7)

where $A_{planned}$ is the planned irrigation area, y_o (t/ha) is the expected crop yield without water deficit, y_f (t/ha) is the final crop yield due to water deficit, and c ($/t) is the price of the crop per ton.

The risky annual rice production given monthly probabilities of (non-)water scarcity are then compared between the reference and the planned irrigation areas (Table 3.2) resulting in the Relative Utility Value, with the Relative Utility Value (RUV) defined as (Equation 3.8):

$$RUV = \frac{(U_i - U_r)}{U_r}$$

(3.8)

where, U_r is the expected annual utility (revenue) obtained with the reference irrigation area from observed river discharge, U_i is the expected annual utility obtained with the planned irrigation area from a single global hydrological model or the ensemble of the global hydrological models (Equation 3.4).

If RUV is equal to zero then U_i and U_r are the same and the simulated surface water availability is equal to the observed.

In this study, we base irrigation area planning on rice production. We use an average price of 298 $/t and an expected rice yield of 10.8 t/ha based on regional statistics (NSW DPI, 2016) to estimate the annual agricultural production. The growing season of rice is from November to February (NSW DPI, 2016).

Table 3.2. Contingence table: Evaluating the expected annual utility of planned irrigation areas relative to the reference irrigation area.

	Using **reference** irrigation area r	Using **planned** irrigation area i
Probability of water scarcity	*Risky annual production* R_{WS_r}	*Higher or lower risky annual production* R_{WS_i}
Probability of **no** water scarcity	*Risky annual production* R_{NWS_r}	*Higher or lower risky annual production* R_{NWS_i}
	Expected annual utility U_r	Higher or lower expected annual utility U_i

3.3 RESULTS

3.3.1 Discharge simulations of global hydrological models

The monthly observed (naturalised) and simulated discharges from the eight global hydrological models are shown in Figure 3.3. In general, discharge simulations show an overall agreement with observations, considering that they have not been locally calibrated for this particular basin. In terms of the KGE, the highest values were obtained with JULES and WATERGAP3 (0.704 and 0.732 respectively) the lowest with HTESSEL (0.195). All models, with the exception of LISFLOOD, underestimated the discharge (negative values of percent bias). Correlation values varies from 0.456 for ORCHIDEE to 0.864 obtained for JULES.

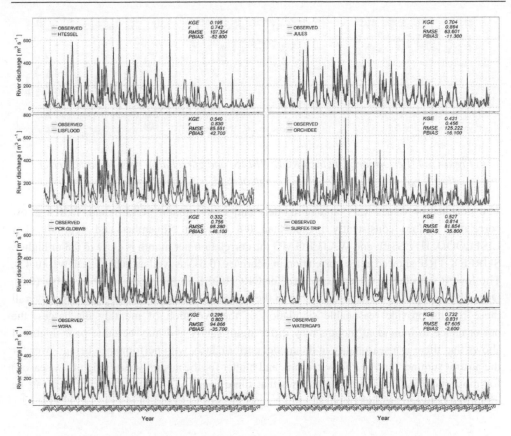

*Figure 3.3. Monthly observed (naturalised) and simulated discharge at Wagga Wagga station
for 30 years (1980-2009).*

The performance metrics for each month are shown in Figure 3.4. As the growth period of rice is between November and February, we evaluate the performance metrics for those months. In terms of the KGE, in November the highest value is obtained with JULES and SURFEX-TRIP with equal values of 0.7. In December, January and February the highest values are obtained with JULES, WATERGAP3 and W3RA, respectively. SURFEX-TRIP, HTESSEL and PCR-GLOBWB underestimate the discharge between November and February with negative values of percent bias. While LISFLOOD and ORCHIDEE overestimate the discharge for those months with the highest PBIAS in February for ORCHIDEE (175%). WATERGAP3 and JULES overestimated the discharge in January and February, and W3RA only in February.

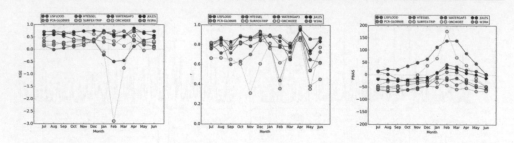

Figure 3.4. Performance metrics KGE, PBIAS, and correlation r for simulated river discharge with single global hydrological models in the Murrumbidgee basin

The correlation values vary among models and each month. JULES and W3RA show correlation values consistently higher than 0.8 between November and February. For LISFLOOD, HTESSEL and WATERGAP3 the correlation per month is close to 0.8 or higher. SURFEX-TRIP showed a low correlation in February (0.61) as well as PCR-GLOBWB (0.45). The lowest correlation value is obtained for ORCHIDEE in the months of November (0.3), December (0.6) and February (0.35), but in the month of January it is as high as 0.81. Overall these results provide reasonable simulation of the evolution of monthly discharges, the performance in individual months may differ significantly and may be quite poor. Figure 4 shows that there is a seasonality in the model performance, with performance being lower for the austral summer (and low flow season), and higher for the austral winter (high flow season) for most models. The figure also shows that there is not a particular ranking that can be identified between the eight models.

3.3.2 Establishing irrigation area estimates

The irrigation area estimates using the simulated water resource availability is established by assessing the water resource availability for the growth period of rice between November and February using a water supply reliability target for each month of 75%. The area is established for the reference discharge, for each of the single global models using the 30-year period, as well as using the ensemble for the six 5-year periods; three 10-year periods, two 15 year periods and the full 30 year period. Results are provided in Table 3. The irrigation areas obtained for all simulations with each single global hydrological model, as well as with the model ensemble are found to be variable larger or smaller than the reference irrigation area (138 km²), with underestimations ranging from 1% to 54% and overestimations ranging from 12% to 306% of the irrigation area (Table 3.3). For the single simulations the closest irrigation area to the reference is obtained with JULES, which agrees with the model performance of JULES in terms of discharge simulation, as the KGE and PBIAS values are among the best results of the single models (see Figure 3.3, Figure 3.4).

The overestimation of the irrigation area represents a higher potential loss in agricultural production, as the probability of water scarcity occurring is higher, while if the irrigation area is underestimated then this represents an opportunity cost in agricultural production, as the irrigation area could have been planned larger, and thus higher agricultural production.

Table 3.3. Irrigation areas obtained using observed river discharge, single and ensemble simulations from global hydrological models.

Irrigation area (km²) using ensemble of global hydrological models for 5, 10, 15 and 30 years

1980-1984	*1985-1989*	*1990-1994*	*1995-1999*	*2000-2004*	*2005-2009*
90	180	289	137	98	79
1980-1989		*1990-1999*		*2000-2009*	
138		193		87	
1980-1994			*1995-2009*		
183			101		
1980-2009					
155					

Irrigation area (km²) using single global hydrological models for 1980-2009 (30 years)

Observed	138
LISFLOOD	562
PCR-GLOBWB	64
HTESSEL	100
SURFEX-TRIP	81
WATERGAP3	281
ORCHIDEE	170
JULES	156
W3RA	184

Irrigation area (km²) using ensemble of global hydrological models for 5, 10, 15 and 30 years

1980-1984	*1985-1989*	*1990-1994*	*1995-1999*	*2000-2004*	*2005-2009*
90	180	289	137	98	79
1980-1989		*1990-1999*		*2000-2009*	
138		193		87	
1980-1994			*1995-2009*		
183			101		
1980-2009					
155					

Irrigation area (km²) using single global hydrological models for 1980-2009 (30 years)

Observed	138
LISFLOOD	562
PCR-GLOBWB	64
HTESSEL	100
SURFEX-TRIP	81
WATERGAP3	281
ORCHIDEE	170
JULES	156
W3RA	184

3.3.3 Probability of water scarcity

The probability of water scarcity using the irrigation area obtained for each single global hydrological model and observed are presented in Figure 3.5 and Figure 3.6. As expected, the probability of water scarcity in February, which is the most critical month, using the irrigation area obtained with the observed river discharge shows a median value equal to 25% and probabilities lower than 25% for the other months.

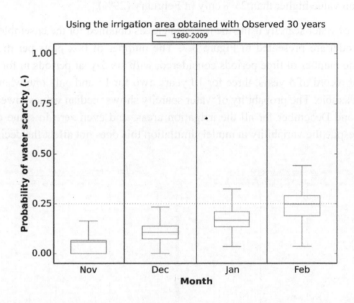

Figure 3.5. Probability of water scarcity using the irrigation area obtained for observed.

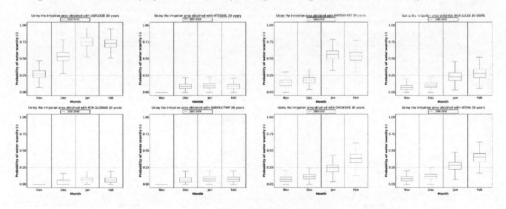

Figure 3.6. Probability of water scarcity using the irrigation area obtained for each single global hydrological model.

The probability of water scarcity shows median values lower than 25% for November and December for all the irrigation areas obtained with the global hydrological models, with exception of LISFLOOD (28% and 53%). For LISFLOOD, the highest probability of water scarcity occurs in January (75%). For HTESSEL, SURFEX-TRIP and PCR-GLOBWB the probability of water scarcity in November is equal to zero and the probability of water scarcity in December, January and February is lower than 25%. This is due to the overestimation of the models during these months. The probability of water scarcity showed median values higher than 25% in January and February for LISFLOOD (75% and 70%), WATERGAP3 (55% and 54%), W3RA (29% and 42%), ORCHIDEE (25% and 40%), with exception of JULES that showed a median value higher than 25% only in February (29%).

The probability of water scarcity using the irrigation areas obtained for the ensembles of global hydrological model are presented in Figure 3.7. The number of box plots per month varies depending on the number of time periods considered, with six 5-year periods in for ensembles with a period of record of 5 years, three for 10 years, two for 15 and only one when using the full 30 years ensemble. The probability of water scarcity shows median values lower than 25% for November and December for all the irrigation areas, and even zero in some cases. This indicates that despite the variability in model simulation this does not affect the decision on the irrigation area.

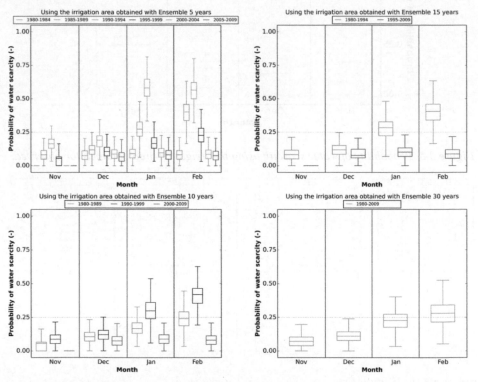

Figure 3.7. Probability of water scarcity using the irrigation area obtained for the ensembles of global hydrological model with 5, 10, 15 and 30 years.

For the ensembles with a length of five years, median values of the probability of water scarcity in January and February vary, with the values above 25% for the second and third of the periods (1985-1989 and 1990-1994), and below 25% for the remaining periods. This reflects that these ten years were relatively wet, resulting in larger irrigation area being selected, though still not as large as in the case of some of the single models. When using ensembles of increasing length, such as for ten years, fifteen, and finally thirty years, the patterns converge to the pattern found using the reference (observed) discharge (Figure 3.5), albeit slightly higher due to the overestimation of the ensemble mean.

3.3.4 Relative Utility Value

The Relative Utility Value using the single global hydrological models against the observed are presented in Figure 3.8 and Figure 3.9. Relative Utility Values (*RUVs*) show median estimates between -0.52 and 2.4 for the single global hydrological models. The Relative Utility Value closest to zero is found for JULES in December (0.12) and the highest value is found for LISFLOOD in February (2.4), followed by WATERGAP3 (0.93). JULES, ORCHIDEE, W3RA, WATERGAP3 and LISFLOOD resulted in positive *RUVs*. Similar values are found in each month for JULES (0.12), ORCHIDEE (0.2) and W3RA (0.3), but for WATERGAP3 and LISFLOOD the *RUV* shows significantly smaller values in December when compared to the other months (results are shown in detail for LISFLOOD in Figure 3.9).

Figure 3.8. Relative Utility Value using seven single global hydrological models against Observed. LISFLOOD is presented in Figure 3.9.

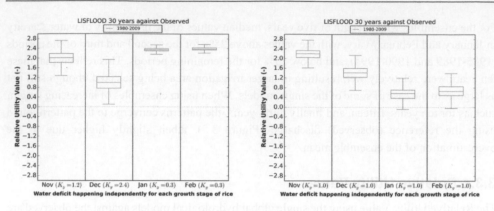

Figure 3.9. Relative Utility Value using LISFLOOD with rice Ky values for each month/growth stage and changed Ky values to 1.0.

LISFLOOD shows the largest statistical spread in *RUV* compared to the other models (Figure 3.8, left panel) with significant value ranges in November (1.3-2.8) and December (0.1-2.0). This occurs because in those months the probability of water scarcity is larger than 25% (which does not occur in the other models, except for WATERGAP), and the yield response factor is higher than 1 (Ky=1.2 in November and Ky=2.4 in December), entailing that the agricultural loss incurred depends not only on the increased occurrence of water scarcity, but also on the sensitivity of the crop to water scarcity occurring. Changing Ky values from 0.3 to 1.0 for January and February results in the *RUV* being closer to zero, but with a larger spread (Figure 3.8, right panel). For WATERGAP3 the largest value range are found between 0.52 and 0.92 in December. Negative values are found for HTESSEL (-0.26), SURFEX-TRIP (-0.4) and PCR-GLOBWB (-0.52), with similar values between months. For the evaluated models, the statistical spread in *RUVs* shows a reduction when obtaining values closer to zero. The largest spread in *RUVs* among the monthly results of every model is found in December, again due to the crop then being more sensitive to water scarcity as defined by the yield response factor.

The Relative Utility Value using the ensembles of global hydrological models against the observed are presented in Figure 3.10. For the ensemble with a period of record of five years, the patterns reflect those found for the probability of water scarcity, though there again the *RUV* is most sensitive to when water scarcity occurs in November and in particular December due to the sensitivity of the crop at this early stage in the growing period. The results show, however, that as the period of record of the ensemble increases, that there is more consistency in *RUV* estimates and the values tend to be closer to zero, which would be the value when using perfect (reference) information.

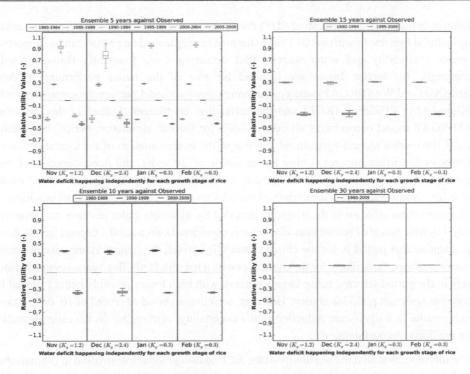

Figure 3.10. Relative Utility Value using ensembles of global hydrological models against Observed with 5, 10, 15 and 30 years.

3.4 DISCUSSION

Results of the Relative Utility Value, which include the irrigation area for different surface water availability estimates from global hydrological models, the monthly probability of water scarcity using these areas, and the potential yield reduction due to water deficit per growth stage of rice, show that using the ensemble of global hydrological models is more beneficial than using single global hydrological models.

In the Murrumbidgee basin we have the good fortune to have a good quality hydrological data to use as reference, but in many areas of the world where expanding irrigated areas is crucial to support food self-sufficiency such in Sub-Saharan Africa (Ittersum et al., 2016) this may not be the case. Water resources estimation may then need to be done with the little information that is available. Given the ubiquitous availability of global model simulations such as provide through re-analysis datasets (Schellekens et al., 2017), we assume that an irrigation area planner without prior hydrological information on surface water availability can choose between any single global hydrological models or an ensemble of models. Additionally the period of record that is used to estimate water resources availability and variability can differ. When using any of the single models considered in this study, the results are shown to be very sensitive to the selection of the model. The possible range of the *RUV* (found here to be -0.52 to 2.4) either a significant opportunity cost (negative *RUV*) due to the planning of too small an irrigation area, or agricultural loss due (positive *RUV*) due to the area being planned large than can be supported by water availability and water scarcity thus occurring more frequently than expected. Interestingly, the largest losses are incurred for two of the better performing models (LISFLOOD and WATERGAP) when considering common model performance statistics such as Kling-Gupta Efficiency (KGE) and the correlation coefficient. Indeed in this case the WATERGAP model outperforms all other models for the full simulation period, but would result in the second highest agricultural loss due to the overestimation of the available water, and thus of the irrigation area. Using the ensemble of 30 years, the overestimation of the irrigation area is minor when compared to the area estimated using the reference observed water availability. Additionally, the uncertainty of the relative utility value is small, thus providing a much more robust estimate of the irrigation area that the available water resource can support. If only a shorter period of record is available (or is considered), then results depend largely how representative that period is for the climate variability, which the irrigation area planner may not know *a-priori*. Particularly for the ensembles with a length of only five years, results depend largely in the period selected being largely normal, with high losses possibly being incurred if an anonymously wet period is chosen. However, selecting a period of record of 10, or 15 years already results in a significant reduction of the uncertainty, converging on the estimate made using the thirty years of data.

The results also show that the sensitivity of the *RUV* of using model information to estimate the irrigation area using the global model data can vary depending on in which month water deficit occurs. While the spread in the estimates of water scarcity is found to be the largest in the months of January and February, the *RUV* may be most sensitive to uncertainty in the estimate of water resources availability in November and particularly December. This is due to different yield reduction values per growth stage of rice, which in this case is the selected crop. In the Murrumbidgee basin, rice has an average growth period of four months with four growth stages starting in November. If water deficit happens in November the yield reduction factor (Ky=1.2) under the same degree of water deficit is four times as high as it is in the later growing stages (Ky=0.3), while in December it is even higher (Ky=2.4). This is the case only if water scarcity occurs in these first months, which happens more frequently for the models that overestimate the available resource. This can be seen for the case of the LISFLOOD model where the probability of water scarcity is higher than 25% the months of November and December, when the Ky is also high, leading to a significant statistical spread in RUV results in those months. The overestimation and uncertainty of the water resources availability is higher in the latter months, but leads to a lower spread in *RUV*. This means that the potential rice production is closer to the reference, but as now the crop is more sensitive to water deficit and the probability of water scarcity is higher than the reference target, the statistical spread of *RUV* is also larger, leading to a higher risk in rice production.

The global hydrological models and ensembles are tested in this case in a highly regulated basin. To allow comparison to the models a naturalised observed discharge was established, which was possible due to the available reference stations. In truly data scarce environments it is recommended to use the proposed methodology in non-regulated basins, were the global model simulations can provide reasonable surface water estimations. If the ensemble of models was severely biased due to for example the operation of reservoirs then these results may change. Although the results found apply for the selected case study only, the real potential of the Expected Annual Utility tool can be expected in data scarce, non-regulated basins around the world to evaluate the benefit of using additional hydrological information (e.g. calibrated hydrological models). In those basins the available hydrological information is low and the potential of using additional hydrological information is higher (Kaune et al., 2017).

3.5 CONCLUSIONS

Global hydrological models are available and can be used globally, but the benefit of using these to support water resources management decisions should be measured locally. We develop a tool that applies a hydro-economic framework to assess the benefit of using an ensemble of global hydrological models for irrigation area planning at the local river basin scale. The ensemble of global hydrological models consists of eight global hydrological and land-surface models; LISFLOOD, WATERGAP3, PCR-GLOBWB, SURFEX-TRIP, HTESSEL, ORCHIDEE, JULES and W3RA. The benefit of using either the ensemble over any one of the single models is evaluated through the Relative Utility Value. This value is composed of the expected annual utility in agricultural production given monthly estimates of the probability of water scarcity. Based on the occurrence of water scarcity at the agreed water reliability threshold an estimate can be made of the irrigation area to be planned. The real occurrence of water scarcity will depend on the true (observed) water resources availability. Additional agricultural losses will be incurred if the area planned is too large, as then water scarcity conditions will occur more frequently, while too small an area will result in a production loss or opportunity cost.

In the Murrumbidgee basin in Australia where the framework was tested, it was found that the performance metrics of the global hydrological models for a time period of 30 years relate to the Relative Utility Value, but *RUV* provides a complete hydro-economic risk assessment for agricultural production depending in which month water scarcity occurs in the developing stage of a crop. This shows that while the uncertainty of water scarcity may be larger in the latter stages of the growing season, the impact of uncertainty is higher in the early stages of the growing season due to the yield being much more sensitive during the earlier stage of the growing season. Moreover, when using the ensemble of global hydrological models a much more robust estimate of the Relative Utility Value, and also for the irrigation area is found. This robustness increases when using an ensemble with a longer period of record. Results show that the tipping point for using a reduced time period ensemble compared to the full 30 years of a single global models in this case is for an ensemble with a period of record of 15 years or longer.

The application of the hydro-economic framework using available ensemble of global hydrological models is shown here to support irrigation area planning in Australia. However, the true application potential of this approach is in other regions where much less in-situ hydrological information may be available.

4

CAN GLOBAL PRECIPITATION DATASETS BENEFIT THE ESTIMATION OF THE AREA TO BE CROPPED?

Based on: *Kaune, A., Werner, M., López López, P., Rodríguez, E., Karimi, P., and de Fraiture, C.: Can global precipitation datasets benefit the estimation of the area to be cropped in irrigated agriculture?, Hydrol. Earth Syst. Sci., 23, 2351-2368, https://doi.org/10.5194/hess-23-2351-2019, 2019.*

Abstract

The area to be cropped in irrigation districts needs to be planned according to the available water resources to avoid agricultural production loss. However, the period of record of local hydro-meteorological data may be short, leading to an incomplete understanding of climate variability and consequent uncertainty in estimating surface water availability for irrigation area planning. In this study we assess the benefit of using global precipitation datasets to improve surface water availability estimates. A reference area that can be irrigated is established using a complete record of thirty years of observed river discharge data. Areas are then determined using simulated river discharges from six local hydrological models forced with in-situ and global precipitation datasets (CHIRPS and MSWEP), each calibrated independently with a sample of five years extracted from the full thirty year record. The utility of establishing the irrigated area based on simulated river discharge simulations is compared against the reference area through a pooled Relative Utility Value. Results show that for all river discharge simulations the benefit of choosing the irrigated area based on the thirty years simulated data is higher compared to using only five years observed discharge data, as the statistical spread of *PRUV* using thirty years is smaller. Hence, it is more beneficial to calibrate a hydrological model using five years of observed river discharge and then extending it with global precipitation data of thirty years as this weighs up against the model uncertainty of the model calibration.

4.1 INTRODUCTION

As water becomes scarce, efficient decision making based on solid information becomes increasingly important (Svendsen, 2005). Solid information on climate variability and climate change are key to adequately estimate the availability of water for human livelihoods, the environment and agricultural development (Kirby et al., 2014a, 2015), especially for irrigated agriculture, which by volume is the largest user of freshwater (de Fraiture and Wichelns, 2010). Available climatological records used for estimation of water resources availability in the irrigation sector are, however, often short (Kaune et al., 2017), and may not be representative of the full distribution of climate variability. This may particularly be so in developing countries, where the need to develop irrigation areas is the greatest, and can lead to sub-optimal decisions, such as the overestimating or underestimating of the area that can be planted. Local authorities deciding on the irrigated area clearly prefer to use the true record of climate variability to estimate the adequate irrigation area to be able to justify their decision based on expected economic benefits, but these records may often be short.

Recent studies show that hydrological information from remote sensing datasets can be effectively used for estimation of surface water availability (Peña-Arancibia et al., 2016), water accounting (Karimi et al., 2013) and to help improve detection of droughts at basin scale (Linés et al., 2017). Combined with local data these datasets can potentially provide improved information to support decisions in irrigated agriculture. Global hydrological models have been used to estimate the river discharge at basin level for the development of irrigated areas and assess the risk of water scarcity (Kaune et al., 2018), and although these show promising results in large basins, the use of a calibrated local hydrological model may be more suitable in smaller basins (López López et al., 2016) as a finer spatial resolution may then be used and local hydrological processes better represented.

Such local models will typically require some level of calibration, and the challenge is to calibrate these when the period of record of the observed data from available in-situ stations is limited. If the period of record is short, then the data may not provide full representation of the true climatic variability, and the water resource estimate will be conditional on whether the available data is from a relatively wet, normal or relatively dry period. This is particularly relevant in climates that are influenced by phenomena such as the El-Niño Southern Oscillation (ENSO).

Using hydrological models forced by a longer period of record from available precipitation datasets may help improve discharge estimates for reliably determining the irrigated area, as the climatic variability can be better represented. However, model uncertainty, as well as the uncertainty of the representativeness of the model given the data used in model calibration will need to be taken into account. Recently, several global precipitation datasets have become available, based on remote sensing as well as re-analysis models, with periods of record spanning thirty plus years. Examples include the CHIRPS precipitation dataset (Funk et al., 2015), which integrates in-situ meteorological data and global earth observations, and the recently developed MSWEP precipitation dataset (Beck et al., 2017b), which integrates in-situ meteorological data, global earth observations and the ERA-Interim re-analysis datasets. Both have been widely used to assess water availability and the risk of water scarcity and drought events (López López et al., 2017; Shukla et al., 2014; Toté et al., 2015; Veldkamp et al., 2015b).

Despite the opportunities these modern datasets offer, they have largely been neglected by the irrigation sector for the estimation of water resources availability and variability (Turral et al., 2010), which relies primarily on in-situ datasets, even when the availability of these datasets is often limited. Assessing the potential benefit of combining data from available in-situ stations, global earth observations and reanalysis datasets to better estimate surface water availability can therefore be of considerable value to irrigation managers.

We hypothesise that the simulated river discharge for a period of record of thirty years using a calibrated local model forced by datasets such as CHIRPS or MSWEP provides more reliable estimates of water resources availability and the area to be irrigated, than when considering the shorter time series of observed discharge that is used to calibrate the model. This is evaluated through an extended version of the hydro-economic Expected Annual Utility framework that determines the value of using each of the different datasets in determining the areas that can be irrigated as a function of the estimated availability of water.

4.2 METHODS

The Pooled Relative Utility Value *PRUV* used in this study is defined as a joined vector of six samples of the Relative Utility Value. This value includes the irrigation areas for river discharge simulations derived using different precipitation datasets, the monthly probability of water scarcity using these areas, and the potential yield reduction due to water deficit for rice. The workflow of this study is shown in Figure 4.1.

4. Can global precipitation datasets benefit the estimation of the area to be cropped?

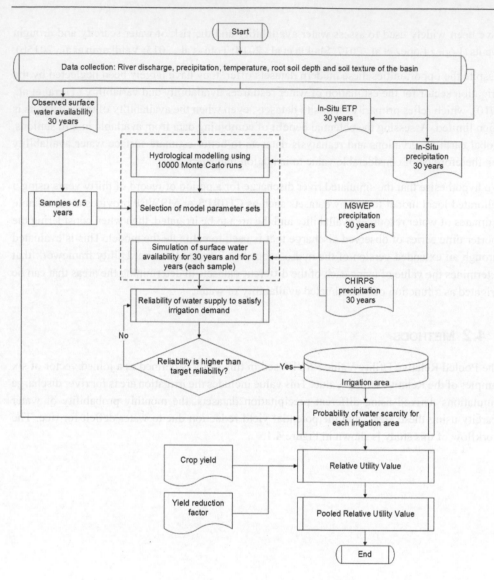

Figure 4.1. Workflow of the study to determine the Pooled Relative Utility Value using different irrigation areas obtained from In-Situ, CHIRPS and MSWEP precipitation datasets.

4.2.1 Coello Irrigation District, Colombia

We apply our analysis to the Coello Irrigation District in Colombia. The Coello Irrigation District is an existing irrigation district located in the upper Magdalena basin, in the Tolima Department, a region subject to considerable climate variability and that is vulnerable to droughts (IDEAM, 2015). The average monthly temperature in the Coello District is 28°C, with maximum daily temperatures reaching 38°C (station 21215080). The reference evapotranspiration is between 137 mm/month in November and 173 mm/month in August with a mean annual evapotranspiration of 1824 mm/year. The irrigation district serves an irrigated flatland of approximately 250 km², comprising mainly of irrigated rice, which is planted continually throughout the year (Urrutia-Cobo, 2006). Rice total growth length is four months with high sensitivity to water deficit at the flowering stage. Resulting yields are between 5.5 and 6.8 t/ha (DANE, 2016; Fedearroz, 2017). The local authority in charge of the water management of the irrigation district (USOCOELLO) reports a gross irrigation demand rate of 0.2 m³/s/km². This is a high demand rate mainly due to low application and conveyance efficiencies, and high evapotranspiration demand.

The water available for irrigation depends on the total discharge of two rivers from neighbouring mountainous basins; the Coello and Cucuana Rivers (Figure 4.2). The Coello basin has an area of 2000 km², and the Coello River has a length of some 112 km starting at 5300 m.a.s.l. and flowing into the Magdalena River at 280 m.a.s.l. with an average flow of 23 m³/s (Vermillion and Garcés-Restrepo, 1996). In the Coello basin the precipitation is bimodal with two peak months in May (186 mm/month) and October (127 mm/month) and two low months in January (50 mm/month) and August (90 mm/month). The mean annual precipitation is 1268 mm/year. The Cucuana basin has similar physical characteristics to the Coello basin. In this research, we focus only on the Coello basin to estimate the surface water availability for irrigation, as the available discharge data from the Cucuana River (Corea Station) is too short for the purpose of our experiment.

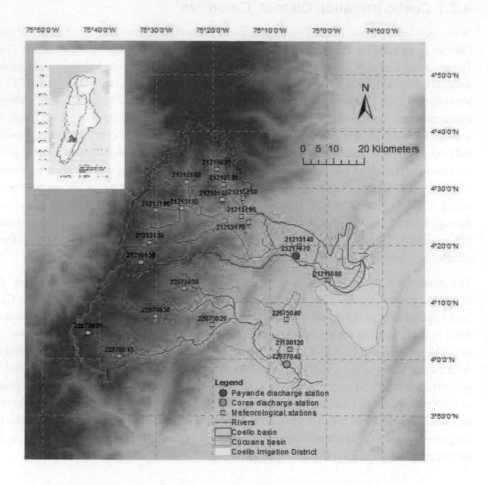

Figure 4.2. Map of the Coello and Cucuana River basins and the Coello irrigation district, and their location in the Magdalena macro-basin in Colombia. The points indicate discharge stations and the squares indicate meteorological stations.

4.2.2 Hydro-meteorological data

In-situ precipitation and temperature data were obtained from the network of meteorological stations operated by the *Instituto de Hidrología, Meteorología y Estudios Ambientales (IDEAM)*, the Colombian hydro-meteorological institute, and interpolated to a gridded dataset with 0.1° spatial and daily temporal resolution for the whole Magdalena-Cauca basin (Rodriguez et al., 2017). The temperature data was used to estimate potential evapotranspiration with the Hargreaves method (Hargreaves, 1994).

Two global precipitation datasets were considered: (i) the Climate Hazards Group InfraRed Precipitation with Station data, CHIRPS (Funk et al., 2015) and (ii) the Multi-Source Weighted-Ensemble Precipitation, MSWEP (Beck et al., 2017b). CHIRPS precipitation is a remotely sensed and ground corrected dataset available globally at 0.05° resolution, while MSWEP precipitation is a merged gauge, satellite and reanalysis dataset available globally at 0.25° resolution. The meteorological stations used in the Coello basin to derive each precipitation product are shown in Table 4.1. A total of 14 stations were used for the in-situ product. For CHIRPS and MSWEP, 7 stations and 3 stations were used, respectively.

4. Can global precipitation datasets benefit the estimation of the area to be cropped?

Table 4.1. Precipitation stations used for each precipitation product (In-Situ, CHIRPS and MSWEP). Positive sign means that the product includes the station. Negative sign means that the product does not include the station.

Station code	Location		Stations used per precipitation product		
	Lat	Lon	In-situ	CHIRPS	MSWEP
21210020	4.56	-75.32	+	+	-
21210030	4.51	-75.30	+	+	+
21210120	4.49	-75.24	+	+	-
21210150	4.28	-75.54	+	+	+
21210160	4.47	-75.30	+	-	-
21210180	4.52	-75.41	+	+	+
21215080	4.23	-75.00	+	+	-
21215100	4.44	-75.42	+	-	-
21215130	4.34	-75.52	+	+	-
21215140	4.33	-75.08	+	-	-
21215160	4.47	-75.24	+	-	-
21215170	4.40	-75.23	+	-	-
21215180	4.42	-75.25	+	-	-
21215190	4.44	-75.50	+	-	-
Total stations			14	7	3

All precipitation, temperature and potential evapotranspiration datasets are available for the 1983-2012 period. A preliminary evaluation of the global precipitation datasets was done. The global precipitation datasets (CHIRPS and MSWEP) were compared against in situ data in the selected basin. The performance indicators KGE, percentage of bias (Pbias) and Pearson correlation (r) were used. The evaluation was done for multi-annual monthly precipitation for the selected 30 year period (Figure 4.3).

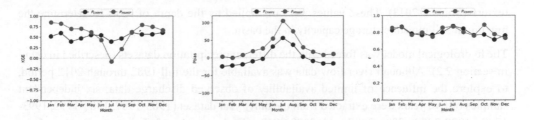

Figure 4.3. KGE, Pbias and r performance metric for monthly CHIRPS and MSWEP precipitation in the Coello basin for 30 years (1983-2012).

KGE results show that MSWEP performs better than CHIRPS from October to May. Only in July, MSWEP performs poorly (KGE=-0.1, Pbias=100%). We cannot discard the use of MSWEP neither of CHIRPS. At this stage, we can recommend the use of each dataset for specific months.

Daily river discharge data for the 1983-2012 period was obtained from the stations operated by IDEAM at the limnigraphic station Payande (21217070) in the Coello River.

4.2.3 Hydrological modelling

The Dynamic Water Balance Model (Zhang et al., 2008), a lumped conceptual hydrological model based on the Budyko framework (Budyko, 1974), was selected to simulate the river discharge in the Coello basin at monthly time scale. The Dynamic Water Balance Model has been applied in several basins around the world (Kaune et al., 2015; Kirby et al., 2014a; Tekleab et al., 2011; Zhang et al., 2008), showing reliable river discharge simulations at monthly time scale. The model has a simple structure without routing, simulating the basin hydrological processes with a reduced number of parameters. There are only four model parameters; basin rainfall retention efficiency α_1 (-), evapotranspiration efficiency α_2 (-), recession constant d (1/month); and maximum soil moisture storage capacity S_{max} (mm). Low (high) values of basin rainfall retention efficiency or evapotranspiration efficiency implies more (less) direct runoff. The recession constant d characterises baseflow, with parameter values ranging between zero and one. The maximum soil moisture storage capacity relates to the root soil depth and soil texture of the basin. As the Coello basin is small, routing processes can be ignored when estimating monthly water availability, which is calculated as the accumulated runoff in the basin upstream of the point of interest.

In this study, surface water availability for irrigation was established as the discharge in the Coello River, considering an environmental flow of 25% from the available water resources. An average maximum soil moisture storage capacity S_{max} of 176 mm was determined for the Coello basin based on the soil texture and the depth of roots in the region. The soil texture and the depth of roots were derived from soil and vegetation maps provided by the *Instituto Geográfico Agustín Codazzi* in Colombia at a scale of 1:500,000. Typical values of the available water storage capacity of the soil in millimetres per meters of depth were used based on the soil texture (Shukla, 2013). These values were multiplied by the depth of roots to determine the maximum soil moisture storage capacity in the basin.

The hydrological model was forced with the different precipitation datasets (described in detail in section 2.2). Although river flow data was available for the full 1982 through 2012 period, to explore the influence of limited availability of observed discharge data, six independent samples of five years were extracted from the thirty year dataset (1983-1987, 1988-1992, 1993-1997, 1998-2002, 2003-2007 and 2008-2012). Each sample of five years was used for calibration of the model parameters (Figure 4.4).

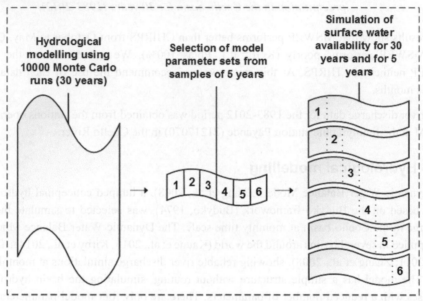

Figure 4.4. Obtaining hydrological model simulations from the six samples of 5 years of observed river discharge.

These samples were extracted as contiguous samples of five years to represent different climatological periods, and were applied to calibrate six sets of models, each using one of the observed discharge samples. Preliminary a Monte Carlo simulation was developed to obtain the full period of samples and then extracting each sample for calibration. 10000 model parameter sets (α_1, α_2, and d with values uniformly distributed between 0 and 1) were generated, and subsequently forced with the full dataset of thirty years of in-situ precipitation data (1983-2012). From this simulation, the five best performing model parameter sets are selected for each of the five year samples based on the comparison of the simulated and observed discharges for the corresponding period. Model performance is measured using the Kling-Gupta Efficiency (KGE) metric (Gupta et al., 2009). This resulted in five calibrated models for six five-year samples, which were used to provide simulated discharge data at the Payande Station for the full thirty year period, forced by each of the three precipitation datasets. Performance metrics of the mean discharge simulation of the five models were calculated separately for each month across the six periods to evaluate the hydrological performance, including KGE, Pearson's correlation coefficient (r), and percent bias (Pbias).

4.2.4 Determining the irrigated area

Similar to Kaune et al. (2018), the area that can be irrigated is determined based on an operational target monthly water supply reliability (R =75%), which means that the monthly demand is met for on average 75% of the years (Equation 4.1). The target reliability depends on local requirements and agreed terms. We selected 75% based on local consultation. The monthly irrigation demand varies depending on the irrigation area, which is the variable to be obtained. A fixed demand rate of 0.2 m³/s/km² was used in the Coello Irrigation District (Kaune et al., 2017).

$$R \leq p\{(Q_a - A_i \cdot r) \geq 0\} \qquad (4.1)$$

where R is the water supply reliability (or probability of non-occurrence of water scarcity, $pr\{NWS\}$), p is the relative frequency, Q_a is the multi-annual monthly surface water availability (considering an environmental flow of 25% from the available water resources), A_i is the planned irrigation area for dataset i (simulated or observed), and r is the demand rate.

The water availability distribution for each calendar month is established using the multi-annual monthly river discharge, which may be obtained either from the observed or the simulated data. Given the small sample size of thirty years, the empirical distribution of water availability is obtained by applying a bootstrap resampling with replacement procedure, with the size of the bootstrap set at 25,000. The bootstrap resampling is applied for each month for the sample of thirty water availability values (multi-annual monthly values). From this sample we randomly draw X values, and leave these out of the dataset. These are then replaced with X values drawn from the remaining values, thus maintaining the same size of the dataset. This process is repeated 25,000 times. The size of the bootstrap is determined iteratively using a progressively increasing sample size until a stable estimate of the empirical distribution is achieved.

A reference irrigated area is established using the empirical distribution derived from the observed monthly river discharges of thirty years (1983-2012). The areas that can be irrigated for each of the six calibrated models is similarly determined but now using the discharge simulations for the full thirty year period. Irrigated areas are additionally obtained for the six five-year samples of observed discharge, and for comparison also using the five year period of simulated discharges for each of the six calibrated model, where the period is commensurate with the period used for calibration. For each irrigation area that is obtained, the real probability of water scarcity is determined using the observed surface water availability (which is also a multi-annual monthly bootstrap resample), and the demand calculated using the estimated area (Equation 4.2).

$$pr\{WS\} = p\{(\widehat{Q_a} - A_i \cdot r) < 0\} \qquad (4.2)$$

where $pr\{WS\}$ is the probability of occurrence of water scarcity, p is the relative frequency, $\widehat{Q_a}$ is the observed surface water availability, A_i is the planned irrigation area obtained from Eq. 1, and r is the demand rate. These probabilities are then used to determine the expected annual utility to evaluate the economic value (Table 4.2).

Table 4.2. Evaluating the expected annual utility using the planned irrigation area from selected river discharge information relative to expected annual utility using the reference irrigation area.

	Using **reference** irrigation area	Using **planned** irrigation area
Monthly probability of water scarcity	*Annual production under probability of water scarcity*	*Higher or lower annual production under probability of water scarcity*
Monthly probability of **no** water scarcity	*Annual production under probability of **no** water scarcity*	*Higher or lower annual production under probability of **no** water scarcity*
	Expected annual utility	Higher or lower expected annual utility

4.2.5 Evaluating the cost of choosing the irrigation area

The cost of choosing the irrigation area was evaluated with an extended version of the hydro-economic framework developed by Kaune et al. (2018) based on the economic utility theory (Neumann and Morgenstern, 1966). The cost is calculated as the opportunity cost when the irrigation area is selected to be too small, or the production loss due to water scarcity when the irrigation areas is selected to be too large. When the area selected is equal to the reference area, then the cost is zero. Similar to Kaune et al. (2018), the Relative Utility Value, RUV is used to compare the expected annual utility between the reference and the irrigated area derived using either the simulated discharge or the shorter five year observed discharge sample (Equation 4.3).

$$RUV = \frac{(U_i - U_r)}{U_r} \tag{4.3}$$

where U_r is the expected annual utility (revenue) obtained with the reference irrigation area from observed river discharge, and U_i is the expected annual utility obtained with any of the irrigation areas obtained from the discharge simulations described in section 2.4.

The expected annual utility U is defined as the expected annual crop production, given monthly probabilities of (non-) water scarcity, and considering a loss in crop production if water scarcity does happen in any one month (Equation 4.4).

$$U = pr\{WS\} \cdot (P_{NWS} - L_{WS}) + (1 - pr\{WS\}) \cdot P_{NWS} \qquad (4.4)$$

where $pr\{WS\}$ is the monthly probability of water scarcity defined in section 2.4; P_{NWS} is the expected annual crop production ($P_{NWS} = c \cdot A_i \cdot y_e$) which includes the irrigation area A_i obtained from section 2.4 and converted into hectares, price of the crop per ton ($/t) and the expected crop yield y_e (t/ha); and L_{WS} is the annual production loss if water scarcity happens in any one month.

For determining the annual production loss L_{WS} an approach is applied where each month corresponds to the growth stage distribution of the crop based on information provided by the Coello Irrigation District. We assume that only rice is grown in the Coello Irrigation District with a growing length of four months sowed over the entire year. The loss in annual rice production L_{WS} due to water scarcity happening in any one month is determined with Equation 4.5:

$$L_{WS} = c \cdot A_i \cdot \Sigma_{month}(y_e - y_a) \qquad (4.5)$$

where y_e is the expected harvested crop yield in a month (t/ha) and y_a is the actual harvested crop yield in a month (t/ha) due to water shortage happening in any one month. Water shortage happening in any one month of the four month crop period will lead to a yield reduction. The actual harvested crop yield y_a obtained, is determined with the FAO water production function in Equation 4.6 (FAO, 2012):

$$\left(1 - \frac{y_a}{y_e}\right) = K_y \left(1 - \frac{ET_a}{ET_p}\right) \qquad (4.6)$$

where K_y is the weighted average yield reduction value per month calculated from established yield reduction factors due to water deficit for each growth stage of rice (FAO, 2012), and the distribution of growth stages as reported by the Coello Irrigation District; ET_a is the actual evapotranspiration and ET_p is the potential evapotranspiration. In our experiment, the actual evaporation is unknown, as this will depend on irrigation scheduling and practice as well as precipitation. As this detail is beyond the scope of this research, we assume the reduction in evapotranspiration $\left(1 - \frac{ET_a}{ET_p}\right)$ to be 20% for the reference irrigation area when water shortage occurs. 20% is selected as this is the evapotranspiration deficit that rice farmers can easily cope with (FAO, 2012). To account for the increased deficit for irrigation areas selected to be larger than the reference area, the evapotranspiration reduction is increased proportionally, assuming the available water is uniformly distributed in the new irrigation area. For irrigation areas selected to be smaller the reduction is decreased proportionally. An average price of 329 $/t and an expected rice yield of 6.8 t/ha based on national statistics (DANE, 2016; Fedearroz, 2017) are used to estimate the expected annual rice production.

If *RUV* is equal to zero, then the expected annual utility obtained with the reference and simulated irrigation areas are the same, and there is thus no cost associated to using the simulated information. A negative *RUV* entails an opportunity cost due to the planning of too small an irrigation area (defined as cost type 1). A positive *RUV* entails an agricultural loss due to the area being planned larger than can be supported by water availability and water shortages thus occurring more frequently than expected (defined as cost type 2). The statistical spread of *RUV* is derived from the bootstrap resample. The spread depends on the probability of water shortage being larger compared to the reference and on the yield response factor, entailing that the production loss incurred depends not only on the increased occurrence of water shortage, but also on the sensitivity of the crop to water deficit.

RUVs are pooled as to give a Pooled Relative Utility Value (*PRUV*) to evaluate the cost of choosing the irrigation area from the six possible irrigation areas obtained for a river discharge simulations. This is done as it is not *a-priori* clear when only five years of observed data are available, from which part of the full climatological record these may be. The Pooled Relative Utility Value (*PRUV*) is a concatenated vector of the Relative Utility Value obtained for each calibration sample (Equation 4.7).

$$PRUV = \{ RUV_1 \ \| \ RUV_2 \ \| \cdots \| \ RUV_6 \} \qquad\qquad (4.7)$$

where *PRUV* is the Pooled Relative Utility Value and RUV_x is the Relative Utility Value for each calibration sample (in this case six samples).

Similar to *RUV*, the *PRUV* is a hydro-economic indicator that can be larger (cost type 2), equal (no cost) or smaller than zero (cost type 1). The statistical spread of *PRUV* encompasses the variability of *RUV* among the six calibration samples. If the statistical spread of *PRUV* is large, then the variability of planned irrigation areas is large among samples. This means that the cost of choosing the irrigation area based on the available information is high. If on the other hand the statistical spread of *PRUV* is small, then the variability of planned irrigation areas is also small and the cost of choosing the irrigation area based on the available information is low.

4.3 RESULTS

4.3.1 Discharge simulations

The monthly observed and the simulated discharges calculated with the different precipitation datasets from the calibration samples are shown in Figure 4.5 (only CHIRPS with the samples for the 1993-1997 and 1998-2002 periods are shown), and in Appendix A (all samples and In-Situ and MSWEP). Discharge simulations change depending on which precipitation dataset is used as forcing and which sample is used to calibrate the hydrological model. In general, however, the mean discharge simulations show an overall agreement with observations.

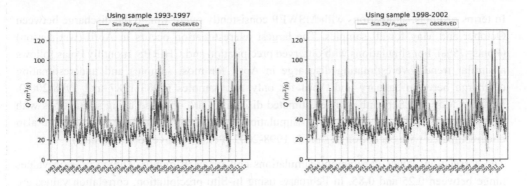

Figure 4.5. Observed and simulated discharge for the Coello River at Payende with 30 years (1983-2012) of CHIRPS precipitation (Sim 30y P_{CHIRPS}) for calibration samples 1993-1997 and 1998-2002, with the sample used to calibrate the model indicated in the header.

The performance metrics for each month are shown in Figure 4.6 and in Appendix A. In all months using the discharge simulations with different precipitation datasets, positive KGE values are obtained with the exception of simulations with MSWEP in April and November, which are both wet season months. In February (dry season) the highest KGE value is obtained using the simulations with observed precipitation (0.75). For all samples in February (dry season), the KGE value is higher for discharge simulations with observed precipitation and CHIRPS than those using MSWEP, with exception of one sample (2008-2012).

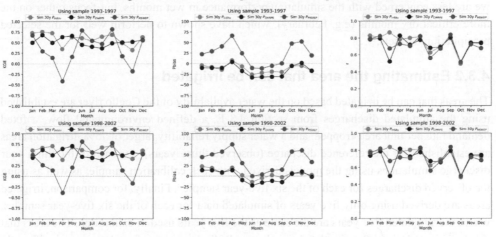

Figure 4.6. KGE, Pbias and r performance metric for simulated river discharge for the complete time period of 30 years (1983-2012) using three different precipitation datasets (In-Situ, CHIRPS and MSWEP) in the Coello basin. Two calibration samples are shown (1993-1997, 1998-2002), with the sample used to calibrate the model indicated in the header.

In terms of Pbias, simulations with MSWEP consistently overestimate the discharge between October and May for all samples. The largest overestimation occurs in April (wet season) (Pbias=75%). For simulations with observed precipitation and CHIRPS, monthly Pbias follows a similar trend, overestimating discharge in April for most samples and underestimating discharge between January and April for only two samples (1983-1987 and 1998-2002). Between May and September underestimated discharges are obtained using simulations with In-Situ and CHIRPS for all samples. Simulations with MSWEP in June and July are also underestimated with exception of sample 1998-2002 (Pbias positive for all months).

The correlation values vary among simulations and for each month. The correlation values range between 0.25 and 0.85. In February, using In-Situ precipitation, correlation values are above 0.6. Simulations with CHIRPS and MSWEP results in correlation values between 0.7 and 0.8 in February. The largest difference between correlations occurs in March (CHIRPS correlation is 0.5, MSWEP correlation is 0.6, and In-Situ correlation is 0.8).

Simulations with In-Situ precipitation and CHIRPS are found to be behave similarly, which is not surprising as CHIRPS uses station corrected data. MSWEP also includes station corrected data, but it is derived in part from the ERA-Interim data which in itself is not good at capturing convective precipitation (Leeuw et al., 2015). This explains the poor simulation performance with MSWEP in April and November as these are wet months in a tropical region with predominant convective precipitation.

As our work is focused on determining the critical irrigation area under monthly water scarcity, we are less concerned with the simulation performance in wet months, but focus rather on the more critical dry months (e.g. February), which have shown to perform well for the selected precipitation datasets.

4.3.2 Estimating the area that can be irrigated

The areas that can be irrigated based on the water availability of the Coello river are established using the simulated discharges from section 4.3.1., a defined environmental flow, a fixed demand rate per unit area cropped, and a water supply reliability target of 75%. Irrigation areas are established for the reference discharge (observed thirty years); for each of the thirty year discharge simulations using the models derived with each calibration sample; as well as using the observed discharges for each of the six five-year samples. Finally, for comparison, irrigated areas are derived using only five years of simulated data for each of the six five-year samples, where the simulated five years are the same as the five years used in calibration. The areas that can be irrigated given the simulated (or observed) discharges are found to vary significantly when compared to the reference irrigated area (which was established as 67.45 km²), with areas ranging from 2% to 40% smaller, to 1% to 69% larger (Table 4.3). In the case of the thirty year simulations, the irrigation areas obtained are found to be always larger than the reference area, with the overestimation ranging from 3% to 69% (Table 4.3), with exception of one sample where an underestimation of 3% when using the observed precipitation is found. The largest estimates of areas that can be irrigated is obtained with MSWEP simulation (69%), which agrees with discharge model performance (section 3.1), as KGE and r values are the lowest for this model, while the bias values are the highest of all the simulations. For simulations with CHIRPS and observed precipitation, Pbias is positive for sample 1998-2002 in the dry months (e.g. February), leading to an overestimation of the irrigation area of 40%. For sample 1993-1997, Pbias is negative for these simulations (close to -10% in February), resulting in a lower estimation of the irrigated area. In this case, an underestimation of 3% in the area is found for the simulation with observed precipitation, but an overestimation of 3% in the area is found for the simulation with CHIRPS. This is related to the difference in variability as the water availability is derived based on the distribution and not on the mean.

Table 4.3. Irrigation areas obtained using different datasets of river discharge information in the Coello basin. The observed river discharge from the complete period of record of 30 years (1983-2012) is the reference information. The irrigation areas are obtained for an agreed water supply reliability of 75% in any one month.

Hydrological information used		Six samples of observed river discharge of 5 years					
		1983-1987	1988-1992	1993-1997	1998-2002	2003-2007	2008-2012
		Size of the irrigation area (km²)					
Observed river discharge	Q_30 years (reference)	67.45					
	Q_5 years	67.93	54.66	65.92	75.18	60.82	65.43
Simulated river discharge	P_In-Situ_30 years	98.97	85.24	65.48	93.93	79.07	84.77
	P_In-Situ_5 years	51.53	47.68	57.76	79.14	42.70	52.48
	P_CHIRPS_30 years	92.38	81.72	69.37	94.56	78.32	80.97
	P_CHIRPS_5 years	47.84	40.20	61.47	84.13	43.13	45.80
	P_MSWEP_30 years	105.90	99.58	86.94	113.97	94.53	98.47
	P_MSWEP_5 years	58.56	58.40	74.17	109.09	55.61	59.69

The areas that can be irrigated that are obtained using the observed discharges for each of the six five year periods show relatively small variation when compared to the reference area, ranging from 19% smaller to 11% larger. The average area of the six five year samples is slightly smaller at 64.99 km^2, just 2.5% smaller than the reference. Conversely, the areas derived using the simulations for each of the five year periods shows that these vary quite considerably, with an overestimation ranging from 10% to 62% and underestimation ranging from 9% to 40% across all precipitation sources. This range is comparable for all three precipitation forcing datasets, indicating that the variability can be attributed primarily to model error, conditional on the five year dataset used in calibration.

4.3.3 Probability of water scarcity

The probability of water scarcity using the irrigation areas obtained for each of the simulated and observed discharges for the five year periods as well as for the reference are shown in Figure 4.7 and Figure 4.8 (samples 1993-1997 and 1998-2002) and in Appendix A (all samples). The probability of water scarcity using the irrigation area obtained using the observed discharges are shown in Figure 4.7. As expected, the probability of water scarcity in February, which is the most critical month, shows a median value equal to 25% and probabilities lower than 25% for the other months when using the irrigation area obtained with the reference discharge (30 years). The spread of the probability of water scarcity indicated by the box-whiskers plot, showing the mean, interquartile range and minimum and maximum, is due to the distribution of the bootstrap, representing the uncertainty in the estimate due to the 30 year period of record.

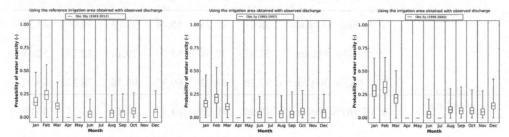

Figure 4.7. Probability of water scarcity using the reference irrigation area obtained with the observed river discharge of 30 years (Obs 30y) and the reference surface water availability. Probability of water scarcity using the irrigation area obtained with the observed river discharge of 5 years (Obs 5y) and the reference surface water availability. Boxplots show the median, interquartile range and minimum-maximum range.

Figure 4.7 similarly shows the probability of water scarcity for irrigation areas obtained using observed discharges for the five year periods; 1993-1997 and 1998-2002 (results for the other four periods included in Appendix A). This shows that for the period 1993-1997, the median value is lower than 25% for all months, while for the period 1998-2002 the median value is higher than 25% for January and February. This reflects the smaller, respectively larger irrigated areas established with each of these datasets. Figure 8 shows the probability of water scarcity for irrigation areas obtained using the discharge simulations of thirty years. The probability of water scarcity in February shows median values higher than 25%, commensurate with the overestimation found in the hydrological model, with the exception of one simulation using observed precipitation, calibrated with the 1993-1997 sample of observed discharge data. Between April and June and in October/November, using the irrigation areas obtained with the discharge simulations, the probability of water scarcity is always found to be lower than 25%, as these are the two wet seasons of the bimodal climate. For all samples, the probability of water scarcity is the highest for the simulations using MSWEP precipitation.

Using the irrigation areas obtained from the simulations calibrated with the 1983-1987 and 1998-2002 samples show higher probabilities of water scarcity for all months when compared to the simulations calibrated with the other samples. This reflects that these years were relatively wet, influencing discharge simulations and resulting in larger irrigation areas being selected. The pattern for sample 1993-1997 is more similar to the pattern found using the reference area found with the 30 years observed discharge.

The probability of water scarcity for irrigated areas obtained with simulated discharges of only five years are shown in Figure 4.8 (again results for the 1993-1997 and 1998-2002 samples are shown, with the remaining four periods provided in Appendix A). The monthly probabilities of water scarcity show large differences between samples. In this case, four out of the six samples do not show median probability of water scarcity higher than 25% for any month, meaning that the irrigation area is underestimated compared to the reference. For the 1998-2002 sample, the probability of water scarcity is the highest with a median probability of water scarcity between 50% and 75% in February.

71

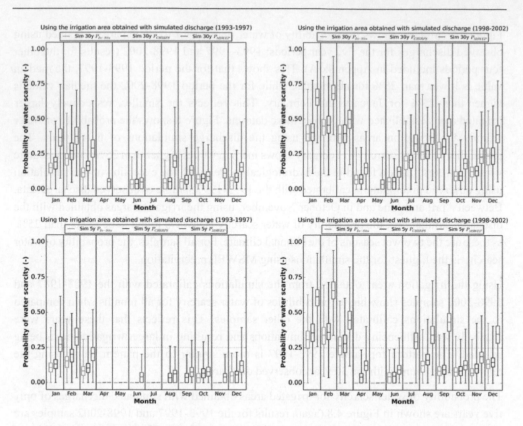

Figure 4.8. Probability of water scarcity using the irrigation area obtained with simulated river discharge information (Sim 30y $P_{In-Situ}$, Sim 30y P_{CHIRPS}, Sim 30y P_{MSWEP};Sim 5y $P_{In-Situ}$, Sim 5y P_{CHIRPS}, Sim 5y P_{MSWEP}) and the reference surface water availability. Boxplots show the median, interquartile range and minimum-maximum range.

4.3.4 Relative Utility Value

The annual expected utility is calculated using the economic return of the rice crop and the estimated yield determined using the irrigated areas established with the simulated discharge information, and the probability of water scarcity in each month for the 30-year period based on the observed discharges. Relative Utility Values are then found through comparing these against the annual expected utility calculated using the reference area and discharge information.

Figure 4.9 shows the Relative Utility Values obtained for areas determined using the five year samples of observed discharge for the 1993-1997 and 1998-2002 periods (again the remaining four periods are provided in Appendix A), with median estimates of -0.02 and 0.11, respectively.

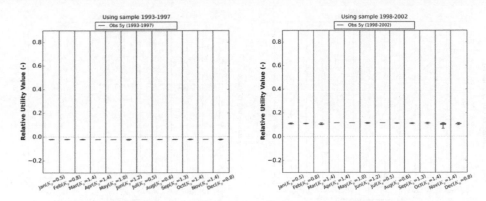

Figure 4.9. Relative Utility Value using observed river discharge of 5 years for water scarcity happening independently in any one month. Ky is the sensitivity of the crop to water deficit. Boxplots show the median, interquartile range and minimum-maximum range.

Relative Utility Values obtained for areas determined using discharge simulations of thirty years are shown in Figure 4.10 (and in Appendix A), with median estimates between -0.03 and 0.65. The Relative Utility Values closest to zero are found for simulations using both the observed and the CHIRPS precipitation datasets, -0.03 and 0.03, respectively, both when using the 1993-1997 sample for model calibration. Of the six samples, this five year period was already noted to be most representative of the whole 30-year period.

4. Can global precipitation datasets benefit the estimation of the area to be cropped?

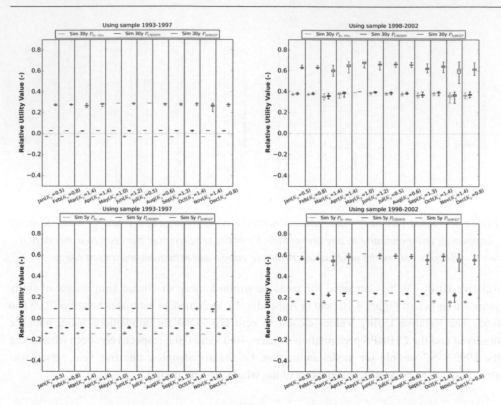

Figure 4.10. Relative Utility Value using simulated river discharge of 30 years and 5 years for water scarcity happening independently in any one month. Ky is the sensitivity of the crop to water deficit. Boxplots show the median, interquartile range and minimum-maximum range.

For all samples, the Relative Utility Values for simulations using the MSWEP dataset are found to be largest, with values between 0.3 and 0.65, indicating a higher production loss due to the higher probability of water scarcity. For simulations using the 30-year observed precipitation, consistent median values between 0.18 and 0.45 are obtained, with the exception of one sample (-0.03). Those obtained with CHIRPS simulations are consistent with those found using the observed precipitation.

The Relative Utility Values obtained using irrigated areas determined with the simulated discharges of only five years (Figure 4.10 and Appendix A), show median estimates between -0.2 and 0.6 (MSWEP), which are larger than the simulations of thirty years. The *RUV* closest to zero are found for simulations with CHIRPS (-0.09), while results for simulations using the observed precipitation show more consistent values closer to zero. In this case, results show more negative *RUVs* for each simulation forcing, and results are less consistent between samples compared to the results obtained with the thirty years.

For the 1993-1997 period, the *RUV* obtained for the irrigated area determined with observed discharge of five years (-0.02) and the *RUV* obtained with for the area determined with simulated discharges of 30 years using either the CHIRPS or the observed precipitation (0.03 and -0.03) are similar and close zero. The extended precipitation period compensates the model uncertainty and results in reliable *RUV* estimates.

A large statistical spread in *RUV* is found in months where the probability of water scarcity is higher than the reference. This is clearly shown for MSWEP simulations, which have the largest estimates of the irrigated area. For months where the probability of water scarcity is lower than the reference, the statistical spread in *RUV* is low. In these cases the statistical spread of *RUV* is a result only of the spread of the reference annual expected utility, resulting from the distribution of the probability of water scarcity. The statistical spread of the *RUV* values is lower when the simulated annual expected utility and the reference annual expected utility are more similar, which means that the *RUV* is closer to zero as shown when using the 1993-1997 sample. An absence of statistical spread for the *RUV* values reflects zero probability of water scarcity in both the simulated and the reference expected annual utility.

Even though the probability of water scarcity is not the highest in November, the statistical spread of the *RUV* is the largest when water scarcity happens in that month. This is due to the high sensitivity of the crop to water deficit (K_y=1.4) in November, which then becomes the determining factor for obtaining a large statistical spread. On the other hand, the smallest statistical spread, or no statistical spread of the *RUV* is found when water scarcity happens in February or May. We select February, May and November as the representative months for further analysis with the Pooled Relative Utility Value.

4.3.5 Pooled Relative Utility Value

The Pooled Relative Utility Value *PRUV* is obtained from the *RUV* values for each of the six samples in section 4.3.4. In Figure 4.11 and Figure 4.12, the *PRUV* results for areas estimated using the observed discharges, and for the simulated discharges for five and thirty years are shown for November, February and May. These are the representative months identified in section 4.3.4, with similar results found for *PRUVs* when water scarcity happens independently in each month.

4. Can global precipitation datasets benefit the estimation of the area to be cropped?

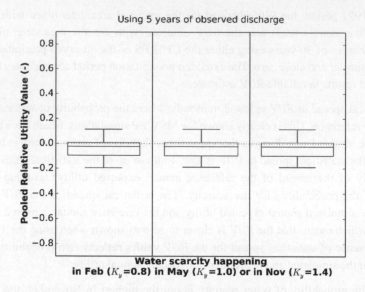

Figure 4.11. Pooled Relative Utility Value using observed river discharge of 5 years for water scarcity happening independently in February, May or November. Ky is the sensitivity of the crop to water deficit. Boxplots show the median, interquartile range and minimum-maximum range.

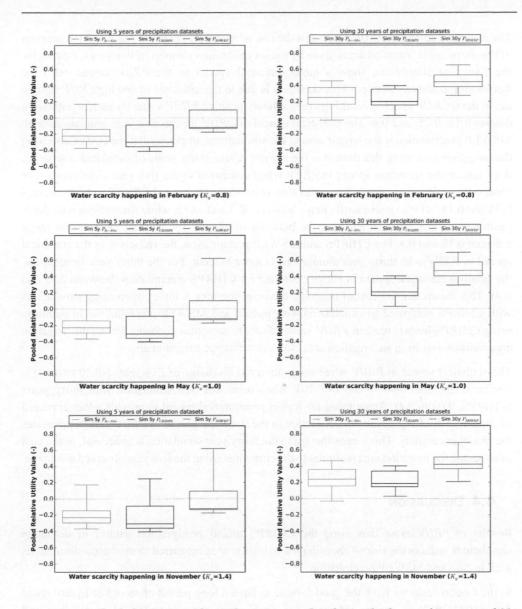

Figure 4.12. Pooled Relative Utility Value using simulated river discharge of 5 years and 30 years for water scarcity happening independently in February, May or November. Ky is the sensitivity of the crop to water deficit. Boxplots show the median, interquartile range and minimum-maximum range.

77

The statistical spread of *PRUV* represents the risk of randomly choosing one irrigation area out of the six possible irrigation areas given by the six calibration samples of five years. Results for the five year simulations show a large statistical spread of the *PRUV* values, with the distribution positively skewed. This skewness is due to the influence of one high *RUV* sample out of the six *RUV* samples, resulting in a maximum positive *PRUV* value for each precipitation dataset 0.18, 0.25, and 0.6. The statistical spread of *PRUV* for the five year simulations with MSWEP precipitation is the largest among the simulations, implying that the cost of choosing the irrigation area using this dataset is the highest. Using thirty years of simulated discharges does reduce the statistical spread in *PRUV* when compared to the five year simulations. For observed precipitation, simulations with five years show the range of *PRUV* to be between -0.38 and 0.18, with an interquartile range between -0.3 and -0.15; while simulations with thirty years show a the range of *PRUV* to be between -0.03 and 0.5, with an interquartile range between 0.18 and 0.4. For CHIRPS and MSWEP precipitation, the reduction in the statistical spread in *PRUV* with thirty year simulations is more evident. For the thirty year simulations, the smallest statistical spread in *PRUVs* is found for CHIRPS precipitation (between 0.03 and 0.4). This means that the cost of choosing the irrigation area is lower when using simulations with CHIRPS compared to simulations with In-Situ and MSWEP precipitation. In addition, using CHIRPS leads to median *PRUV* values closer to zero, thus choosing among the irrigation area samples results in an irrigation area closer to reference irrigation area.

The statistical spread in *PRUV* when using observed discharge of five years (-0.20 to 0.12) is similar to the statistical spread in *PRUV* when using the best simulation with thirty years (CHIRPS, 0.03 to 0.4). Again using the longer precipitation record to provide a longer record of simulated discharge results in a reduction in the statistical spread in *PRUV* and compensates the model uncertainty. This means that using the thirty year simulation is beneficial, as the cost of choosing the irrigation area is similar to the cost when using the five year observed discharge.

4.4 DISCUSSION

Results of *PRUV* show that using the CHIRPS global precipitation dataset in discharge simulations reduces the risk of choosing the irrigation area compared to discharge simulations with In-Situ and MSWEP precipitation.

In the Coello basin we have the good fortune to have a long period of record of hydrological data (1983-2012) to use as reference for establishing the climatological availability and variability of the available water resource. This may not be the case in other basins. Water resources estimation may then need to be done with the limited information that is available. To help understand the risk of estimating the available water resources when only limited information is available, we used observed discharge with a shorter period of record (five years) to calibrate a local hydrological model and apply this to obtain simulated discharge with a longer period of record (30 years) either using a precipitation dataset based on observed data (Rodriguez et al., 2017) or a global precipitation dataset, including CHIRPS and MSWEP (Beck et al., 2017b; Funk et al., 2015). We establish six samples of five years to calibrate the parameters of a hydrological model, and simulate six possible discharges of thirty years to imitate a setting where information about how representative the short record of available observed discharge is not known *a-priori*. For each sample the annual expected utility is determined including the monthly probability of (non-) water scarcity using different irrigation areas from different discharge simulations and the annual crop production with water scarcity (not-) happening in a month. Positive and negative Relative Utility Values were found with different discharge simulations. Positive values indicate a crop production loss due to unexpected water scarcity for too large an irrigation area being planned. Negative values indicates an opportunity cost due to the planning of too small an irrigation area.

Results show that the *RUV* varies depending on in which month water scarcity happens. While the spread in the estimates of probability of water scarcity is found to be the largest in the month of February, the spread in *RUV* is larger when water scarcity happens in November. This is due to the difference in sensitivity of the crop yield to water deficit, depending on the growing stage of the crop. In the Coello basin, rice has an average growing length of four months and is sowed during the entire year. This means that if water scarcity does happen in a particular month, four different growth stages will be affected each with a different yield reduction factor resulting in an average yield reduction value. If water scarcity happens in November the average yield reduction value from the four growing stages of rice is 1.4. This means that the average yield reduction in November under an equal degree of water deficit is 1.75 times higher than in February (K_y=0.8). If water scarcity occurs in February, even though the probability of water scarcity is higher than the reference 25%, the statistical spread in *RUV* is low due to the low average yield reduction value. Using different sources of river discharge information to estimate the irrigation area will indeed change the estimates of the monthly probability of water scarcity changing the *RUV* values. However, the impact on annual production may be low if water scarcity occurs in the month where the sensitivity of the crop to water deficit is low. Reducing agricultural production losses depends not only on using adequate river discharge information to estimate the irrigation area, but also on adequate planning of the crop stage distribution.

For an irrigated area selected based on the estimate of water availability using simulated discharges, a decision maker takes an additional risk due to not knowing *a-priori* how representative the data used for calibrating the model is of climatic variability. This is why we introduce the Pooled Relative Utility Value *PRUV* in order to evaluate the risk of choosing an irrigation area derived from different river discharge simulations. If the statistical spread of *PRUV* is low (high), then the cost incurred by choosing an irrigated area based on the results of the simulations is equally low (high). The Pooled Relative Utility Value results using the global precipitation CHIRPS showed a lower cost in choosing the irrigation area compared to *PRUV* results using both a dataset based on observed precipitation, as well as the MSWEP global precipitation dataset. This would suggest that the CHIRPS precipitation should be used instead of both observed and MSWEP precipitation when determining the surface water availability for irrigation area planning to avoid the risk of agricultural production loss due to a poorly chosen irrigated area that can be supported based on water availability. This is not a general conclusion, as it is closely related to how representative the precipitation dataset used is of the true precipitation amount and variability in the basin. The CHIRPS dataset does include observed data (Funk et al., 2015), which is similar as that used in our study to establish the In-Situ precipitation dataset. In that sense, it is also an interpolated dataset, but with additional information from the satellite. This may well provide additional detail on the variability of precipitation in a tropical mountainous basin such as the Coello.

Important to mention is the fact that CHIRPS and MSWEP are gauge corrected. This would mean that they would both be expected to perform quite well. However, the datasets used to correct each product may differ. That is way we compare the amount of stations used in the Coello basin for each of the precipitation products (In-Situ, CHIRPS and MSWEP). This is helpful for discussing the *PRUV* results. Even though the amount of stations used is lower for correction in the CHIRPS product (7 stations) compared to the number of stations used in the In-Situ product (14), the results indicate that the satellite information included in CHIRPS still provides a reasonable representation of the basin precipitation. For the MSWEP product the only three stations are used for correction, resulting in a poorer representation of the rainfall in the basin. In summary, the basin precipitation dataset derived from CHIRPS for the Coello basin is better than the MSWEP. The higher resolution of the CHIRPS dataset when compared to that of MSWEP no doubt also contributes in this medium sized, mountainous basin. The poorer comparison of the MSWEP data we found not to be immediately obvious when evaluating the precipitation data using common indicators (e.g. KGE, bias), but was only found when evaluating the hydrological information for determining the irrigated area.

Interestingly, the performance of the model using the observed precipitation dataset is similar to that of the model using the CHIRPS precipitation dataset when considering common model performance statistics such as Kling-Gupta Efficiency (KGE), percentage bias (Pbias) and the correlation coefficient (r). The MSWEP product includes reanalysis datasets additional to observed and satellite datasets, but instead of providing a benefit its local application in this small to medium sized basin in Colombia has a negative influence on the representation of the climate variability. In that sense, our results match with previous research where the performance of reanalysis datasets in regions dominated by tropical warm rain processes is not the best (Beck et al., 2017b) attributed in part to the poor prediction of convective precipitation (Leeuw et al., 2015). Moreover, the 0.25° resolution might be too coarse to represent the spatial variability in the basin which undervalues the potential use of these datasets in such conditions. The further development of the MSWEP dataset, including an improved resolution of 0.1 degrees may increase its value for applications such as that explored in this research (Beck et al., 2018). Additionally, the period of record of consistent data for datasets such as MSWEP, but also of CHIRPS continues to increase. For now, in Colombia, where the availability of observed precipitation is reasonable (IDEAM, 2015; Kaune et al., 2017), the CHIRPS datasets appears provide the best estimates of surface water availability in basins larger than 2000 km² for determining the irrigation area. The benefit of CHIRPS, MSWEP and other such global precipitation dataset can be evaluated in other case studies around the world using the proposed framework. Certainly there is a new opportunity for the irrigation sector in using modern hydro-meteorological data and information to improve water allocation decisions considering the economic impacts of uncertainty in those datasets.

An irrigation manager may be reluctant to use simulated information instead of the observed until it is proven that the additional period of record of precipitation from for example global datasets compensates the uncertainty of the use of a hydrological model. In that sense, *PRUV* results provide evidence that using discharge simulations with thirty year precipitation (CHIRPS) is equivalent in using observed discharge of five years as the risk of choosing the irrigation area is similar. As the period record of datasets such as CHIRPS increases, this risk will be expected to reduce further.

Using a longer period of record of observed discharges will help make better estimates of the irrigated area that can be supported by the available water resources, but when the availability or quality of observed discharge is limited, extending the period of record using model based discharge simulations provide an alternative to estimate the area to be cropped. The results of the model used in the Coello basin also show that the overestimation or underestimation of the planned irrigation area depends in part on the model bias, particularly in the ability of the calibrated model to provide reliable simulations for low flow periods, which are the most critical in this application. In the case presented here, we use a very simple model structure, and using a simulated discharges from an enhanced model structure can be explored to obtain more accurate results.

4.5 CONCLUSION

We apply an extended hydro-economic framework to assess the benefit of using global precipitation datasets in surface water availability estimates to reduce the risk of choosing the area that can be irrigated with available water resources based on limited available information. We estimate irrigation areas using observed river discharge with a period of record of thirty years (reference), and simulated river discharges from a hydrological model forced with In-Situ and global precipitation datasets (CHIRPS and MSWEP). The hydrological model is calibrated using independent observed river discharge samples of five years extracted from the reference time period of thirty years to emulate a data scarce environment, as well as the uncertainty of the available data with a short period of record being fully representative climate variability. The Relative Utility Value of using a particular dataset is determined based on the reference and simulated annual expected utility, which includes the monthly probability of (non-) water scarcity using the irrigation areas obtained and the annual crop production with water scarcity (not-) happening in a month. The monthly probability of water scarcity will depend on the true (reference observed) water resources availability. Additional production losses are incurred if the irrigation area planned is too large, as then water scarcity conditions will occur more frequently (cost type 2), while too small an area will result in an opportunity cost (cost type 1). The production loss also depends on how sensitive the crop is to water deficit in a particular month. The benefit of using either the In-Situ, CHIRPS or MSWEP datasets in reducing the cost of choosing the irrigation area, irrespective of the available sample of observed data used in calibrating the model, is evaluated through a Pooled Relative Utility Value, a joined estimate of the Relative Utility Value of the samples of five years.

4. Can global precipitation datasets benefit the estimation of the area to be cropped?

In the Coello basin in Colombia where the framework was applied, it was found that while the performance metrics of the discharge simulations relate to the Relative Utility Value, the Pooled Relative Utility Value provides a complete hydro-economic indicator to assess the risk of choosing the irrigation area based on observed or simulated discharge data. We find that for the Coello basin, the CHIRPS precipitation dataset is more beneficial than In-Situ or MSWEP precipitation, as the risk of choosing the irrigation area is lower due to a better estimate of climate variability. For all precipitation datasets evaluated, using a dataset with a length of thirty years leads to a lower risk when compared to using a length of only five years. The risk of choosing the irrigation area based on discharge simulations with thirty years of CHIRPS precipitation is found to be similar to using the observed discharge of five years. Hence, extending the period of record using an extended precipitation dataset to provide a longer record of discharge simulations (from five to thirty years) compensates the model uncertainty of the model calibration.

In the Coello basin, the global precipitation data CHIRPS is recommended instead of global precipitation data from the MSWEP dataset for estimating surface water availability to support the planning of irrigation areas. This dataset provides a good representation of the climatic variability in this medium sized tropical basin, in part due to the correction of the dataset using observed station data. While the performance of the available global precipitation datasets would need to be evaluated, the application of the extended hydro-economic framework using global precipitation datasets to force a locally calibrated hydrological model is shown here to support decisions on adequate selection of irrigated areas in Colombia, and can be applied in data scarce basins around the world. Ensuring the use of adequate hydrological information for the estimation of surface water availability will promote improved decisions for irrigation area planning and prevent economic losses.

5

THE BENEFIT OF USING AN ENSEMBLE OF SEASONAL STREAMFLOW FORECASTS

Based on: *Kaune A., Chowdhury F., Werner M., Bennett J., 2019. The benefit of using an ensemble of seasonsal streamflow forecasts in water allocation decisions. Hydrol. Earth Syst. Sci. (In review).*

Abstract

The area to be cropped in irrigation districts needs to be planned according to the allocated water, which in turn is a function of the available water resource. However, conservative estimates of the available resource in rivers and reservoirs may lead to unnecessary curtailments in allocated water due to conservative estimates of future (in) flows. Water allocations may be revised as the season progresses, though inconsistency in allocation is undesirable to farmers as they may then not be able to use that water, thus leading to an opportunity cost in agricultural production. In this study we assess the benefit of using reservoir inflow estimates derived from seasonal forecast datasets to improve water allocation decisions. A feedback loop between simulated reservoir storage and emulated water allocations to General Security (water allocated to irrigating) was developed to evaluate two seasonal reservoir inflow forecast datasets (POAMA and ESP), derived from the Forecast Guided Stochastic Scenarios (FoGSS), a 12 month seasonal ensemble forecast in Australia. We evaluate the approach in the Murrumbidgee basin, comparing water allocations obtained with an expected reservoir inflow from FoGSS against the allocations obtained with an expected reservoir inflow from a conservative estimate based on climatology (as currently used by the basin authority), as well as against those obtained using observed inflows (perfect information). The inconsistency in allocated water is evaluated by determining the total changes in allocated water made every 15 days from the initial allocation at the start of the water year to the end of the irrigation season, including both downward and upward revisions of allocations. Results show that the inconsistency due to upward revisions in allocated water is lower when using the forecast datasets (POAMA and

ESP) compared to the conservative inflow estimates (reference) which is beneficial to the planning of cropping areas by farmers. Over confidence can, however, lead to an increase in undesirable downward revisions. This is more evident for dry years than for wet years. Even though biases are found in inflow predictions, the accuracy of the available water estimates using the forecast ensemble improves progressively during the water year, especially one and a half months before the start of the cropping season in November, providing additional time for farmers to make key decision on planting.

5.1 INTRODUCTION

Allocating water is the process of sharing the available water among claimants over a period of time (Hellegers and Leflaive, 2015; Le Quesne et al., 2007). Basin authorities are responsible to allocate water among different users; including agriculture, cities, industry and the environment. The available water in rivers and reservoirs, and the demand placed on it, may vary over time and space due to climate variability, climate change and population growth. Hence, allocating the available water poses a challenge for decision makers, especially in increasing drought and water scarcity conditions.

Basin authorities can allocate water following a demand-based approach. This consists of revising or reviewing the expected water demand in the basin at the start of each water year and allocating the required volume (Linés et al., 2018). Over the water year the initial allocation might decrease depending on the availability of water in reservoirs and from upstream catchments. In other basins the water availability is initially reviewed before allocating the entitled allocation. Allocation of water to meet the entitlements of the license holders is based on the estimate of the available water, which is made using the observed stock in the reservoirs at the time of making the decision, as well as the expected inflow.

In Australia, the water allocation process is governed by clear water policy and regulations at basin level, such as defined in the Murray-Darling Basin Plan (Australian government, 2008). The Murray-Darling river system is highly regulated, especially in the basins located in New South Wales (Ribbons, 2009). Predicting the inflow into the reservoirs is key to adequately allocate water, especially for allocation to irrigated agriculture. However, conservatively low estimates of the expected inflow based on climatology are currently used at the beginning of the water year to estimate the available water for the coming season. As the irrigation season progresses, the estimate of available water may be revised. Given the conservative initial estimate the revision is typically upwards, with consequent upward revisions in the water allocated during the year. For most years the allocated water is set too low at the beginning and is then progressively increased until the 'real' estimate of available water is reached and the 'real' allocation can be established. Ideally, water users would like to know their 'real' water allocation at the beginning of the water year. This is especially so for irrigators that depend on accurate and timely water allocation to choose which crop to plant and to decide on the area to be cropped, allowing them to maximise the benefit of the water that their license entitles them to.

The use of seasonal forecasts of reservoir inflows may be beneficial to support water allocation decisions through providing better and earlier estimates of the available water. The potential for farmers to benefit from seasonal forecasts will, however, depend on i) how well the climate can be predicted, ii) how much this information helps in the actual decision process and iii) how much it contributes to reducing negative impacts (Hansen, 2002). Most studies focus on evaluating the benefit of forecasts by determining the skill, often based on forecast results, observed data and a benchmark prediction (Pappenberger et al., 2015). Though this provides relevant insight to the (relative) quality of forecast, it may say little of the benefit to users through improved decisions. The use of seasonal forecasts to support decisions has been addressed in several settings. Winsemius et al. (2014) assess the predictability of meteorological indicators in a changing climate and show how skilful forecasts can support rain-fed agriculture. Shukla et al. (2014) developed and implemented a seasonal agricultural drought forecast system for East Africa which shows to perform well in drought years. Crochemore et al. (2016) assess the performance of seasonal streamflow forecasts to a reservoir with standard indicators of forecast skill such as reliability, sharpness, and accuracy. Anghileri et al. (2016) evaluate the performance of Ensemble Streamflow Prediction (ESP) with the performance based on climatology and perfect forecasts. Turner et al. (2017), in addition to the usual forecast skill assessment, include the performance gain in reservoir operation, benchmarking penalty costs when using forecast against using perfect and actual information. Boucher et al. (2012) apply an ensemble streamflow forecast for determining its value in supporting hydropower generation. However, the potential enhancement to water allocation decisions in irrigated agriculture that are informed by seasonal forecasts has been little studied. A complete assessment of the added value of seasonal forecasts can allow basin authorities to explore the opportunities seasonal forecasts provide to improve their operational decisions, and reduce potential losses to agricultural by improving water allocation estimates.

In this study, we develop and test a water allocation framework to assess the value of using a seasonal forecast to predict the inflow into the reservoirs in the regulated Murrumbidgee basin in Australia. We hypothesise that water allocation decisions can be improved when informed by seasonal forecasts of reservoir inflows. The inconsistency in water allocation decisions during the water year is introduced as a measure of the value of using the forecast in the allocation process under the premonition that if water allocated to farmers changes little during the season that they can then make maximum benefit of their allocation. If the inconsistency in water allocation decisions using the forecast is lower than when using the currently used conservative estimate of water availability then there is value in using the forecast. In addition to the water allocation estimated at the start of the water year, key decision dates for the cropping season are evaluated to determine the benefit to the farmers in supporting the decisions they need to make.

5.2 METHODS

5.2.1 Water allocation process in the Murrumbidgee basin, Australia

The regulated Murrumbidgee River basin (84000 km²) was selected to evaluate the benefit of using forecasted reservoir inflow (Figure 5.1). In this basin, two major water storages, the Burrinjuck and Blowering reservoirs provide the required resource for the water allocation process. The New South Wales Office of Water (basin authority) announces the water allocation for different users in the basin starting July 1st of each year based on the available water in storage, the expected reservoir inflows for the next 12 months, and water requirements downstream. The inflow to the Blowering reservoir depends on both the discharge release from an upstream hydropower system (Snowy Hydro Scheme) and the natural runoff, while the inflow to the Burrinjuck reservoir depends mainly on natural runoff. Currently, a conservative estimate of $2.33 \times 10^8 \, m^3/year$ (based on climatology) is used to determine the expected reservoir inflow for the next 12 months.

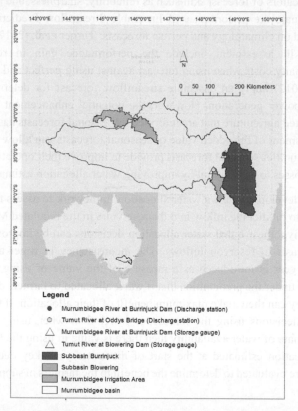

Figure 5.1. Map of the Murrumbidgee River basin in Australia. Location of the Murrumbidgee Irriagtion Area, subbasins for the Blowering and Burrinjuck reservoirs, discharge stations and storage gauges.

In the Murrumbidgee basin, the water year runs from July 1st through to June 30th of the next calendar year, while the cropping season for annual (irrigated) crops is between November 1st and the end of February. Each water user (e.g. irrigation, urban and environment) holds a water entitlement (water share), which is a license to extract water from the basin. Depending on the available water, the basin authority announces a fraction of the total volume of that entitlement that will be met in a year, which is defined as the water allocation (Green, 2011). Water is allocated to different water users according to established priorities. In water abundant conditions each user gets their agreed full entitlement. When water shortage occurs, the highest priority is to satisfy human water consumption and the lowest priority is to satisfy the irrigation demand of annual crops. This is reflected in the type of license users pay for, with irrigators holding a High Security license having a higher priority to water over irrigators holding a General Security license. Normally, irrigators growing annual crops hold a General Security license, though it is the annual crops that require the highest volume of water of the system. Once the water has been allocated, users can decide how they would like to employ the resources available to them. A maximum of 30% of the volume entitlement can be carried over from one year to another. This provides certain flexibility to the water users as they can hold a certain volume of their allocation in the storages and make it available for the next year.

During the water year the basin authority may revise the initial water allocation and announce a new allocation for each user depending on the then available water storage volume and the estimate of the remaining inflow to the end of the water year. . In the currently established regulations the water allocation cannot be decreased, but only maintained or increased, unless exceptional circumstances dictate. This process generates expectations among the irrigators about the amount of water volume they are going to get, especially for the General Security (GS) license holders. To illustrate the process, Figure 5.2 shows the recorded water allocation decisions made in the 2016-2017 water year. On July 1st 2016, the initial water allocation for General Security (GS) license holders was established at 40%. Two weeks later, the water allocation was increased from 40% to 52%. Subsequent revisions over the next four months resulted in a final water allocation of 100% at the start of the cropping season (November), which was subsequently maintained until the end of the season.

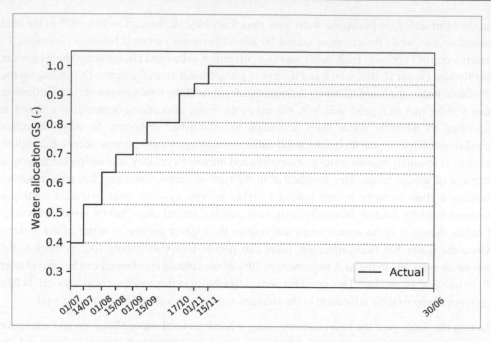

Figure 5.2. Actual water allocation for General Security, GS in the Murrumbidgee basin for the water year 2016. Dotted lines in the water allocation curve are used to show that the water allocation is an annual water volume, the estimate of which changes during the water year.

Certainly, irrigators are aware that the allocated water can be higher in the announcements made closer to the start of the cropping season. This means that farmers may well act on the expectation that the allocation is typically revised upwards, taking the risk of a certain water volume finally being allocated. We argue that this risk affects decision making among the farmers and their ability of correctly planning the area to be cropped. Farmers more adverse to risk may tend towards cropping according to the conservative water allocation, leading to potential losses (opportunity cost), while less risk adverse farmers may face losses in yield if the final allocation they expect is not met. Ideally, water users would like to be informed about the true water allocation as early as possible to better plan their activities. This means that the ideal allocation scenario is when the water allocation is set at the start, and then remains constant during the water year.

5.2.2 Data and information

Data and information from the Murrumbidgee basin was collected from the online repositories of the New South Wales Office of Water and the Australian Bureau of Meteorology (BoM). Actual water allocation for the different users (e.g. General Security) and the observed daily inflow into the reservoirs were obtained for the period 2011 to 2016 (www.water.nsw.gov.au). The current water allocation policy was introduced in 2011, with allocations prior to that date following a different policy. For the 1982-2009 period, actual daily inflow into the two reservoirs was back-calculated using observed daily outflows and observed storage. Discharge data from gauging station 410008 (Murrumbidgee river downstream of Burrinjuck dam) and gauging station 410073 (Tumut river at Oddys bridge) were used to obtain the daily outflow from Burrinjuck and Blowering reservoirs, respectively. Daily reservoir storage volumes were obtained for each reservoir from station 410131 (Murrumbidgee river at Burrinjuck dam – storage gauge) and station 410102 (Tumut river at Blowering dam – storage gauge) (see Figure 5.1).

The forecasted datasets for determining the expected reservoir inflow were obtained from an experimental streamflow forecasting system called Forecast Guided Stochastic Scenarios (FoGSS), available for the time period 1982-2009 (Bennett et al., 2017). Two different datasets from FoGSS were used. One dataset included precipitation and sea-surface temperature (SST) predictions from the POAMA M2.4 seasonal climate forecasting system (Hudson et al., 2013; Marshall et al., 2014). Another dataset is based on resampled historical precipitation, referred to as Extended Streamflow Prediction (ESP) (Day, 1985). A more detailed description of FoGSS is provided in section 5.2.3.

To determine the reservoir inflow, the 10304 km² upstream basin for the Burrinjuck reservoir was considered. The reservoir inflow into the Blowering reservoir is based on the upstream discharge release from the Snowy Hydro Scheme and the natural discharge from a downstream station in the Goobarragandra river at Lacmalac, with an influence basin area of 668 km² (Figure 5.3). Snowy Hydro discharge release was obtained through back-calculating the observed outflow and observed storage data of Blowering reservoir (1982-2009). The inflow from Snowy Hydro into the Blowering reservoir is taken as observed and not subject to the forecast. This means that the forecasts and calculated runoff only applies to the inflow to Burrinjuck reservoir and the Goobarragandra River.

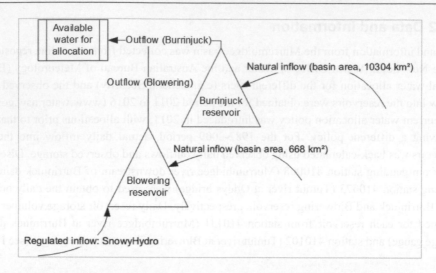

Figure 5.3. Inflows and outflows in the Burrinjuck and Blowering reservoirs in the Murrumbdigee basin for determining the available water for allocation.

5.2.3 Forecast Guided Stochastic Scenarios (FoGSS)

The streamflow forecasting system "Forecast Guided Stochastic Scenarios" (FoGSS) is an experimental forecasting system, which has been developed and tested for the Australian continent (Bennett et al., 2016, 2017; Turner et al., 2017). FoGSS is an ensemble streamflow forecast in the form of monthly time series for a 12-month forecast horizon. As forecast skill declines with lead time, FoGSS is designed to nudge forecasts towards climatology. FoGSS post-processes climate forecasts, either derived from ESP or POAMA to force a monthly hydrological model (Wapaba model). ESP is an ensemble of seasonal precipitation forecasts and an ensemble seasonal forecast based on a coupled ocean-atmosphere general circulation model (CGCM). Calibration, bridging and merging is used to correct biases and remove noise from the CGCM forecasts. The uncertainty in the hydrological model is constrained by an error model which includes, residual normalisation and variance homogenisation, conditional bias-correction and the application of an autoregressive model to improve forecast accuracy and propagate uncertainty through the forecast. Full details of FoGSS can be obtained in Bennett et al. (2016, 2017).

5.2.4 Developing the water allocation decision model

The water allocation decision making process developed in the Murrumbidgee basin consists of a feedback loop between the simulated available water resource and the emulated water allocation decision for the different users. Figure 5.4 schematically shows the decision model to available water resource to the different users. All time dependent variables are from day t to the end of the water season. Users include, in order of priority; Environmental Water (EW= $0.60 \times 10^8 \ m^3/year$), Towns (TD= $0.85 \times 10^8 \ m^3/year$), High Security (HS=$3.60 \times 10^8 \ m^3/year$), Irrigation Conveyance (IC=$3.76 \times 10^8 \ m^3/year$), and General Security (GS=$18.9 \times 10^8 \ m^3/year$).

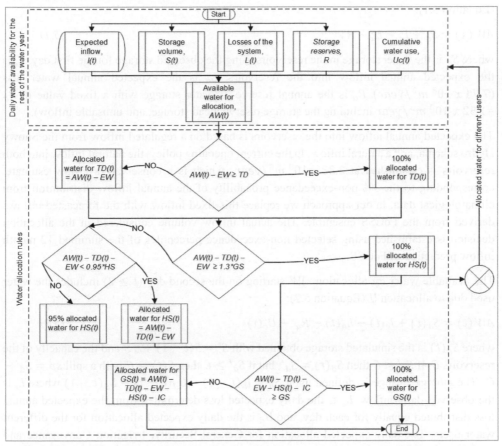

Figure 5.4. Water allocation framework used to simulate the daily water availability for the rest of the water year and emulate the allocated water for different users based on allocation rules in the Murrumbidgee basin. The input variable is the expected inflow into the reservoirs.

The available water is determined at a daily time step and the water allocation decision is emulated for selected announcement dates. The available water is determined considering the storage volume in the reservoirs, expected reservoir inflow, storage reserves and water losses. The expected inflow into reservoirs is the input variable, which feeds into the established water balance to determine the available water for allocation. Water allocation decisions are emulated for the different users from which the water allocation for GS is derived. As water is released from the reservoirs due to the water allocation process a new water availability estimate is determined for the next time step.

5.2.5 Determining the available water for water allocation

The available water for allocation on the first day $(t = 1)$ of the water years is defined as (Equation 5.1):

$$AW(1) = S_d + I_a - L_a - R_a \qquad (5.1)$$

where S_d is the water storage in the reservoirs using the observed storage for the first day, I_a is the expected annual inflow into the reservoirs, L_a is the expected annual water loss $(7.71 \times 10^8 \ m^3/year)$, R_a is the annual reserve of water storage with a fixed value of R_a $=1.52 \times 10^8 \ m^3/year$ including the storage reserve, dead storage, and unusable inflow).

The expected annual inflow into the reservoirs is based on a regulated inflow from the Snowy Hydro schema, and a natural inflow. In the current operating policy, the natural inflow into both reservoirs is established at $2.33 \times 10^8 \ m^3/year$, which is a conservative inflow estimate, corresponding to the 3% non-exceedance probability of the annual inflow distribution from climatological data. In our approach we replace this fixed inflow with the forecasted inflows derived from the FoGSS ensemble. The actual inflow volume considered in the allocation decision is established using selected non-exceedence percentiles of the summed 12 month inflow prediction of FoGSS.

The available water for allocation, AW starting on the second day $(t \geq 2)$ includes the water used due to allocation U (Equation 5.2):

$$AW(t) = S_d(t) + I_e(t) - L_e(t) - R_a + U_c(t) \qquad (5.2)$$

where $S_d(t)$ is the simulated storage obtained with $S' = S(t-1) + \Delta S$ and the capacity of the reservoirs C. If $S_d' < C$ then $S_d(t) = S_d'$, but if $S_d' \geq C$ then $S(t) = C$ with a spill $sp = S_d' - C$. The storage change is defined as $\Delta S_d = I_o(t-1) - U_d(t-1) - L_d(t-1)$ where I_o is the observed daily inflow, L_d is the daily expected loss determined from the expected annual loss distributed equally for each day, and U_d is the daily expected allocation for the different water users. The daily expected allocation for General Security, Irrigation Conveyance and High Security and Towns is determined with a daily release ratio multiplied by the water allocation for each user. For General Security, Irrigation Conveyance and High Security the daily release ratio is based on the monthly irrigation requirements (Table 5.1). For Towns the water allocation is distributed equally for each day. The water allocation for each user is obtained based on the daily available water from the water balance and the allocation rules explained in detail in the next paragraph. $I_e(t)$ is the expected inflow for the remaining days of the water year defined as $I_e(t) = I_e(t-1) - I_d(t-1)$, where I_d is the daily expected inflow determined from an established daily inflow fraction of the average annual observed inflow on day $t - 1$. $L_e(t)$ is the expected loss for the remaining days of the water year defined as $L_e(t) = L_e(t - 1) - L_d(t - 1)$, where L_d is the daily expected loss to the previous day. $U_c(t)$ is the cumulative water use due to allocation defined as $U_c(t) = U_c(t - 1) + U_d(t - 1)$, where U_d is the daily expected allocation for the different water users obtained with the daily release ratio and the emulated water allocation decision.

Table 5.1. Monthly irrigation requirements in the Murrumbidgee basin used in the water allocation model. Values are based on estimations from S. Khan et al., 2004; Shahbaz Khan, Tariq, Yuanlai, & Blackwell, 2006.

Month	Average Inflow Volume (GL)	Average irrigation requirements			
		Monthly ratio (%)	Daily ratio (%)	Volume (GL)	Monthly ratio (%)	Daily ratio (%)
July	470	15%	0.50%	3	0.21%	0.01%
Aug	495	16%	0.52%	26	1.84%	0.06%
Sep	420	14%	0.46%	81	5.76%	0.19%
Oct	401	13%	0.42%	210	15.05%	0.49%
Nov	230	8%	0.25%	245	17.51%	0.58%
Dec	140	5%	0.15%	246	17.63%	0.57%
Jan	110	4%	0.12%	289	20.68%	0.67%
Feb	80	3%	0.09%	219	15.68%	0.56%
Mar	110	4%	0.12%	40	2.89%	0.09%
Apr	120	4%	0.13%	19	1.38%	0.05%
May	170	6%	0.18%	9	0.64%	0.02%
Jun	315	10%	0.34%	10	0.73%	0.02%
Annual	3061	100%		1398	100%	

The water allocation decision is emulated each day, but the final allocated water is presented only for announcement dates starting on July 1st and then for each 15 days to the end of the season. In order to emulate the water allocation decision the framework includes the available water, the agreed entitlement for each water user and the priority rules. The volume is established stepwise for each user following the priority of water use. The allocated water for Towns, stock, domestic and basic right (TD) was set at 100% of the entitlement. Only if the difference between the available water for allocation and the environmental water is lower than $0.85 \times 10^8 \ m^3/year$ (the entitlement of TD), then the allocated water for TD is the difference of AW and EW. The procedure for allocating water to High Security (HS) is presented in Figure 4. It includes the use of the available water AW and the allocated water for TD and EW to obtain three possible outcomes for the allocated water for HS. Depending on availability, water allocation for HS can be 100% or 95% of the entitlement, or the difference of the AW, TD and EW. The allocated water for General Security is determined by using the remaining water volume left after the High Security procedure, subtracting the amount lost to Irrigation

Conveyance ($3.76 \times 10^8 \; m^3/year$). If that difference is lower than the General Security entitlement, then the allocated water for GS is lower than 100%.

The decision model is tested against the recorded allocation decisions for the years from 2011 to 2016 as the current water allocation policy was introduced in 2004 (Horne, 2016).

5.2.6 Evaluating water allocation decisions

To evaluate the water allocation decisions made during the water year a method to quantify the inconsistency in allocated water was applied. The inconsistency (I) can occur due to either upward or downward revisions of the allocated water volume during the water year. An upward revision is when the allocated water at time step t (WA_t) is larger than the allocated water at time step $t-1$ (WA_{t-1}). Hence, the inconsistency in allocated water due to upward revisions (I^+) is the sum of the difference between allocated water at time step t (WA_t) and the allocated water at time step $t-1$ (WA_{t-1}) with that condition:

$$\forall \; WA_t > WA_{t-1} \Rightarrow I^+ = \sum_{t=1}^{n} WA_t - WA_{t-1} \qquad (5.3)$$

A downward revision occurs when the allocated water at time step t (WA_t) is lower than the allocated water at time step $t-1$ (WA_{t-1}). Hence, the inconsistency in allocated water due to downward revisions (I^-) is the sum of the absolute difference between allocated water at time step t (WA_t) and the allocated water at time step $t-1$ (WA_{t-1}) with that condition:

$$\forall \; WA_t < WA_{t-1} \Rightarrow I^- = \sum_{t=1}^{n} |WA_t - WA_{t-1}| \qquad (5.4)$$

A constant allocated water from the beginning until the end of the water year implies zero inconsistency. This would imply that the expected inflow estimates are perfect, and the total water allocation is correctly determined at the start of the season. This water allocation WA_p was derived by applying the observed inflows in the decision model. We separate dry and wet years according to the allocated water obtained with the observed inflow. The years with allocated water equal to 100% of the entitlement considered as wet years, and the years with allocated water lower than 100% are dry years. Average water allocation decisions for dry and wet years were obtained at each time step. The Root Mean Square Difference($RMSD$) was used to evaluate the allocated water obtained with the expected inflows WA_i against the allocated water obtained with the observed inflows (perfect information) WA_p at each time step t for selected years y (dry or wet years) (Equation 5.5).

$$RMSD = \sqrt{\frac{\sum_{y=1}^{m}(WA_i - WA_p)^2_y}{m}} \qquad (5.5)$$

5.3 RESULTS

5.3.1 Emulating historical water allocation decisions

A preliminary calibration of the allocation framework was developed for the 2011-2016 water years using the conservative inflow estimate ($2.33 \times 10^8 \; m^3/year$), and the recorded allocation decisions. The main calibration parameter is the allocation use reduction factor, which determines the percentage of the water allocated to them that users decide to use, with the remainder being reserved for carry-over to the next year. In reality, this factor varies between users as well as between years, and may be influenced by a variety of factors, including many that do not depend on water availability. We simplify this by considering a bulk allocation use reduction factor across all General Security users and also consider this to be equal across years. The allocation use reduction factor derived for the 2011-2016 period is then assumed to also hold for evaluating the FoGSS datasets for the 1982-2009 water years, for which recorded allocation decisions are not available as a different policy for the allocation of water to the different users was then in place. The allocation use reduction factor was established as 78% to obtain similar simulated and actual carry over volumes between water years, as well as the simulated storage in the reservoirs at the end of each water year (itself a function of the carry-over volume which remains in the reservoir). Figure 5.5 shows the simulated water storage in the reservoirs, as well as the observed and simulated carry-over volumes. For the years 2011 to 2016, the emulated water allocation decisions for General Security (GS) are shown in Figure 5.6, and compared to the actual allocation decisions recorded in those years. Two simulations are shown. In the first, the initial storage condition of the reservoirs (day 1 of the 2011 water year) was set equal to the actual water storage followed by continuous daily simulations as in equation 2 for the full six year period. In the second, the water level in the reservoirs is reset to be equal to the observed reservoir level at the start of each year (Simulated Nudged). These simulations show that across the six years volume differences range from 1% to 30% of the actual water storage at the start of each of the water years. Derived emulated water allocation decisions for GS show an underestimation of the allocation compared to the actual volume for most years, especially for the 2014 water year. These differences occur because a constant factor is used (78%) to simulate the carry-over between water years. Results from the Simulated Nudged show how the daily water storage simulations and the water allocation for GS would behave using the actual water storage information (including the actual carryover volume), by nudging the simulated storage levels to the observed at the start of each water year (Figure 5.5). The daily water storage simulations (Figure 5.5) and water allocation to GS are now closer to the actual values, especially for the 2014 water year. However, for both simulations, the emulated decisions show a similar trend when compared to the actual decisions. Upward revisions as well as where the water allocation remains constant (no revisions) occur at the corresponding announcement dates, and although results are slightly biased in volume, the emulation of allocated water decisions does follow the trend of the actual decisions.

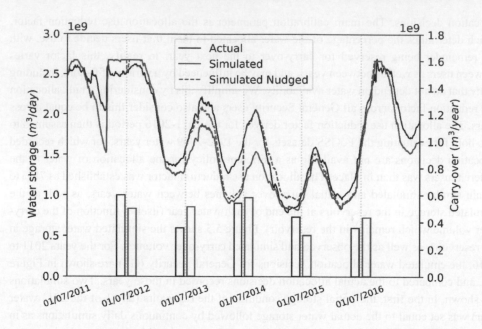

Figure 5.5. Actual and simulated reservoir storage and carry-over volumes in Blowering and Burrinjuck reservoirs (2011-2016)

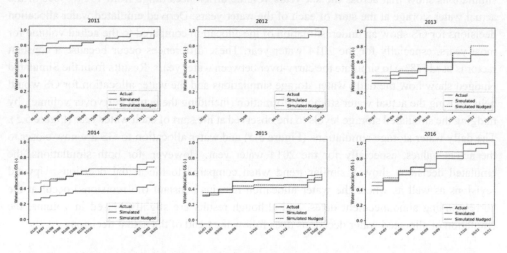

Figure 5.6. Actual and emulated water allocation GS for each year (2011-2016) using conservative inflow. Simulated Nudged is the simulated water allocation GS, but using the actual storage on July 1ˢᵗ

For the 1982-2009 water years, testing our hypothesis on the carryover is not available. We set up the water allocation framework to simulate the water storage continuously over this time period, assuming a constant allocation use reduction factor of 0.78 between water years. For the purpose of testing our hypothesis, the framework is useful as we simulate the available water for allocation and compare results between the use different reservoir inflow information to inform water availability, including the seasonal forecast datasets (FoGSS).

5.3.2 Skill of seasonal predictions of available water

Prior to applying the FoGSS forecasts to emulating the water allocation decisions, the skill of the seasonal predictions of available water is evaluated. In Figure 5.7 and Figure 5.8 rank histograms are shown for both the POAMA and ESP datasets for the forecast ensemble of 1000 members (FoGSS) of the expected inflows into the Burrinjuck reservoir. The expected inflow shown is the total inflow from the forecast month through to the end of the water year, which is the information on the available water that is required to inform the water allocation decision. Rank histograms for expected inflows into the Blowering reservoir show a similar pattern and are available in the Appendix B. The rank histogram shows the frequency of the rank of the observed in the ensemble, with a well calibrated ensemble showing a uniform distribution (Wilks, 2011). For easier interpretation the ensemble is pooled into five classes. The light grey bar shows the frequency of the observed rank being higher (or lower) than the highest (lowest) forecast value in the ensemble.

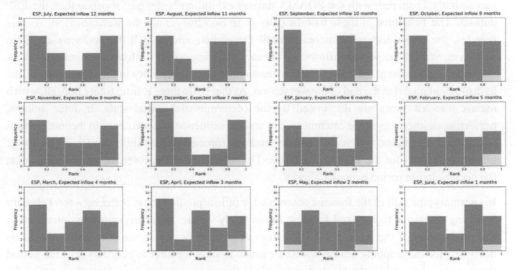

Figure 5.7. Rank histogram using ESP datasets from FoGSS (1982-2009) for expected inflow in the next n months (Starting July) in the Burrinjuck reservoir.

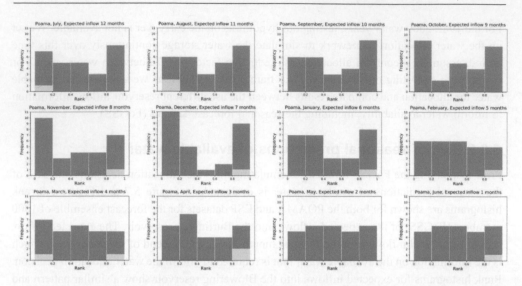

Figure 5.8. Rank histogram using Poama datasets from FoGSS (1982-2009) for expected inflow in the next n months (Starting July) in the Burrinjuck reservoir.

The rank histograms of the expected inflows to end of season for forecasts made in the months from July until January show that the ensemble is under-dispersed as the distribution is increasingly U-shaped, with the POAMA dataset exhibiting better performance than the ESP dataset. The first three of these months (July to to September) are the wetter season. As this recedes, the under-dispersion increases until December, after which the performance again improves, with the expected inflows in February showing a near uniform distribution. Likely the under-dispersion at the end of the wet season is due to uncertainty in the initial state of the hydrological model at the end of the wet season. At that point there is little future rainfall, which means forecast flows are dominated by the (deterministic) initial state. In later months, performance improves as the catchments dry and the influence of uncertainty in the initial states established during the wet season recedes and the forecast results show progressively lower biases, especially for the POAMA datasets. The reliability of the forecast again decreases as the accumulation period becomes shorter.

In summary, the skill of the forecast ensemble for inflow predictions is better between February and June compared to July and January. In our study we are, however, primarily interested in the predictions of the expected inflow from July to February to support water allocation decisions for the cropping season which are made from November to February. For the period that we are interested in, the forecast ensemble is shown to contain biases. How these affect the water allocation decisions, and if using the forecast ensemble leads to better estimates of available water compared to the currently used conservative estimate based on climatology is evaluated by using the ESP and POAMA forecasts to inform the water allocation decisions.

5.3.3 Water allocation using seasonal forecast datasets

Water allocation decisions for General Security (GS) were emulated for the 1982-2009 period using four datasets of expected inflows to the reservoir to determine water availability: (i) observed inflow (considered as perfect information); (ii) the conservative inflow (or reference information as currently used by the decision maker); (iii) the FoGSS seasonal forecast based on POAMA, and (iv) the FoGSS seasonal forecast based on ESP. Water allocations using the perfect information and the reference information provide the benchmarks against which the decisions made informed by the ensemble forecasts are compared. For each of the two ensemble forecasts, two setups were tested. In the first, the inflow prediction at the beginning of the water year is obtained from the ensemble forecast made on July 1st, and this is then maintained for the next 12 months (non-updating FoGSS set up). This means that the forecast of the available water that is established on July 1st is maintained to be the same during the full water year. This was done to mimic the current procedure used by the basin authority when using the conservative inflow estimate based on climatology. In the second set-up that was tested, the full potential of using the ensemble forecast is explored. The water availability estimate to the end of season is now updated each month using the FoGSS forecasted inflows determined from the forecast made at the start of that month. In determining the expected water availability from the FoGSS forecasted inflows for both set-ups, different non-exceedance percentiles of the forecast ensemble are used to provide the expected availability of water for allocation, starting with the 1st non-exceedance percentile (commensurate with a very conservative estimate of water availability), and increasing this to the 50th non-exceedance percentile (commensurate with the ensemble median).

5.3.4 Using one prediction at the beginning of the water year.

Water allocation decisions were emulated using the inflow prediction obtained from the seasonal forecast at the beginning of the water year, and then not updated for the next 12 months (non-updating FoGSS set up). Figure 5.9 shows the water allocation decisions to General Security (GS) for a selected wet year (1998) and a selected dry year (2006) using the 1st, 5th, 10th, 25th and 50th non-exceedance percentile of the ESP and POAMA ensemble forecast datasets. The blue line shows the allocation decisions made using the reference conservative inflow estimate, while the red line shows the allocation established using perfect information.

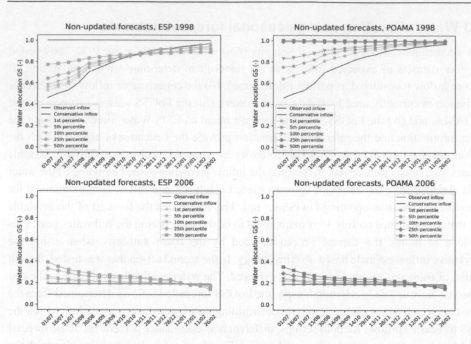

Figure 5.9. Water allocation GS for one wet year (1998) and one dry year (2006) using one inflow prediction at the beginning of the water year for the next 12months (Non-updated forecasts)

For the dry year (2006), the results show that the water allocation decisions using the forecast ensemble are similar to the decisions obtained with the conservative inflow for the 1st percentile. This makes sense as the water allocation using the lowest percentiles is the water allocation that matches best with the water allocation based on the conservative inflow. It is interesting to note that the water allocation based on perfect information is even lower than when using either the conservative inflow or the 1st percentile. This is due to 2006 being the driest year on record (Dreverman, 2013), with observed inflows below the 1st percentile. For the water allocations obtained with the 1st percentile, as well as with the conservative inflow there are no downwards revisions during the water year. However, for increasing percentiles, the number of downward revisions increases as the initial estimate of available water at the start of the year becomes increasingly over confident. For this dry year, the POAMA and ESP forecasts exhibit broadly similar behaviour.

In the wet year (1998), the initial allocation based on the 1st percentile is higher than that using the conservative estimate, particularly for the POAMA dataset. For increasing percentiles, the water allocation decision approach those using perfect information. For 2006, the observed water availability was well above the requirement to fulfil 100% of the allocation to General Security. No downward revisions are found, even for the higher percentiles.

These results provide an initial comparison between using the conservative inflow and forecasted inflow for determining the available water for allocation. However, for the full potential of the seasonal forecast to be evaluated, inflow predictions are updated monthly during the water year as new FoGSS forecast become available.

5.3.5 Updating the inflow prediction every month

In Figure 5.10 and Figure 5.11, the water allocation decisions are shown using the 1st, 5th, 10th, 25th and 50th non-exceedance percentiles of the ESP and POAMA forecast datasets, with monthly updated inflow predictions. Results for the dry years are shown in Figure 5.10, with those for wet years shown in Figure 5.11. Of the 28 years evaluated, 13 are considered as dry (1982, 1997 and 1999-2009, the latter period constituting the millennium drought) and 15 as wet (1983-1996 and 1998). Wet years are taken to be those years when the final water allocation to the General Security attains 100% of the concession by the end of the season. For the dry years we show results for three years (1982, 2003, and 2006), with the results for the remaining years provided in the Appendix B. The selected three years have different levels of water allocation based on perfect allocation (70%, 55%, and 10% of the full concession), reflecting increasingly severe drought conditions. The reference water allocation based on the conservative climatological estimate is again shown in red. For the wet years we equally show results for three wet years (1988, 1995, and 1998), with the three different patterns of the reference water allocation corresponding to increasingly dry conditions. Results for the remaining years are again provided in the Appendix B.

For both dry and wet years, results found with POAMA and ESP differ only slightly in magnitude, and the trend during the water year is similar. For most of the years evaluated, results obtained using the forecast ensembles show that the derived water allocation decisions tend towards those established with the perfect water allocation (which was established using the observed inflow). For all wet years and many of the dry years, the water allocations using the forecast ensemble are generally closer to those made using the perfect water allocation compared to those made with the reference water allocation. In the wet years results closest to the forecast are obtained with the higher forecast percentile (less conservative estimate), while for the dry years the lower percentiles provide the best results.

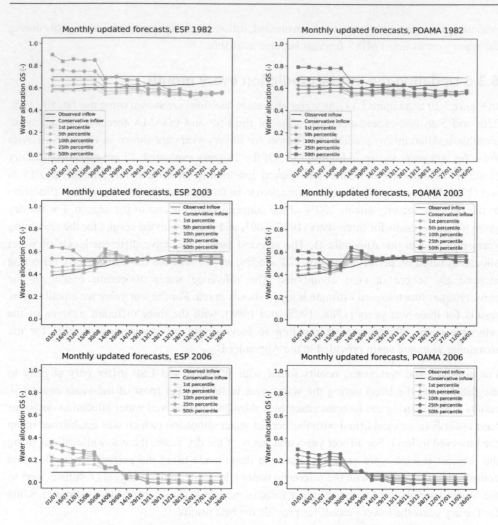

Figure 5.10. Water allocation GS for selected dry years (1982, 2003 and 2006) using new inflow predictions every month (Monthly updated forecasts).

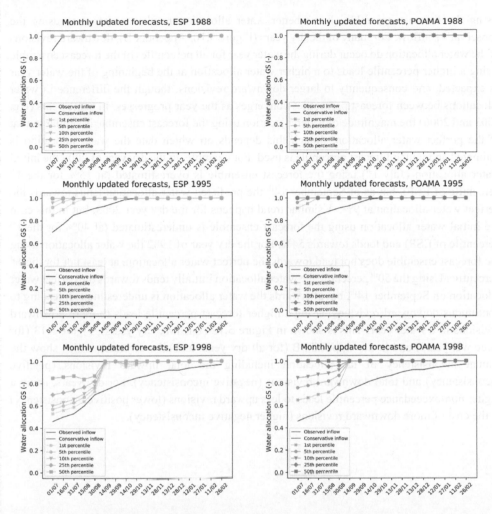

Figure 5.11. Water allocation GS for selected wet years (1988, 1995 and 1998) using new inflow predictions every month (Monthly updated forecasts).

For some of the dry years (1982 and 2006 in Figure 5.10), the water allocation using the conservative inflow initially overestimates the "real" water allocation. The water allocation is then revised downwards as the season progresses. In 2006, the water allocation based on the conservative inflow is 10% higher than the perfect water allocation during the entire year Observed inflows for the 2006 water year were the lowest on record (Dreverman, 2013). Similar behaviour is found in 2007 and 2008, though initial storages at the start of these years in these dry years were so low that allocations never exceeded 0%. The overestimation at the start of the season may well be attributed to a wet bias in the forecast for these dry years, with the more conservative forecasts (1% - 10%) thus providing the best estimate of actual inflows. In other years, the water allocation using the conservative inflow is lower than the perfect water allocation at the beginning of the water year and progressively increases (e.g.1982 and 2003).

Using the forecast ensemble shows better water allocation results compared to using the conservative inflow. For many of the dry years (Figure 5.10, Appendix B), downward revisions of the water allocation do occur during the water year for all percentiles of the forecast ensemble. Using a higher percentile leads to a higher water allocation at the beginning of the water year as expected, and consequently to larger downward revisions, though the difference in water allocations between forecast percentiles converges as the year progresses. In some years (e.g. 2003 and 2006) the magnitude of water allocation using the forecast ensemble is similar to that of the perfect water allocation, though this depends on which date the water allocation is estimated and which forecast percentile is used. For example, in the dry year of 2006 the initial water allocation (July 1st) using the forecast ensemble is overestimated (at 15% for the 1st percentile of ESP), but as of September 14th the predicted water allocation tends towards the perfect water allocation at 9%. A similar trend happens for the dry year 2003, but in this case the initial water allocation using the forecast ensemble is underestimated (at 40% for the 1st percentile of ESP) and tends towards 54%. For the dry year of 1982 the water allocation using the forecast ensemble does not tend towards the perfect water allocation, at least not for the 1st percentile. Using the 50th percentile the water allocation initially tends towards the perfect water allocation on September 14th, but afterwards the water allocation is underestimated, leading to continuous downward revisions. Using a higher forecast percentile leads to more downward revisions in a water year. This is shown in Figure 5.12 (for three dry years), Figure 5.13 (for three wet years) and in the Appendix B (for all dry years and all wet years), which show the annual inconsistency of the allocation including the total upward revisions (positive inconsistency) and total downward revisions (negative inconsistency). For dry years, using a higher non-exceedance percentile leads to less upward revisions (lower positive inconsistency) at the cost of more downward revisions (higher negative inconsistency).

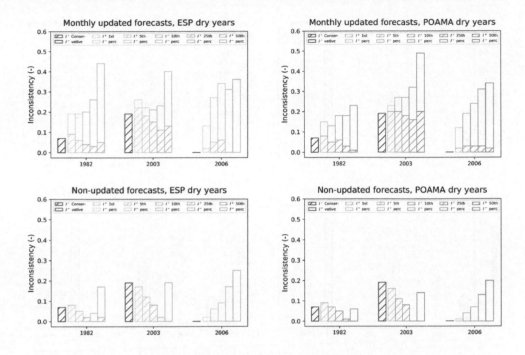

Figure 5.12. Inconsistency for selected dry years (1982, 2003 and 2006) using new inflow predictions every month (Monthly updated forecasts) and one inflow prediction for the next 12 months (Non-updated forecasts).

105

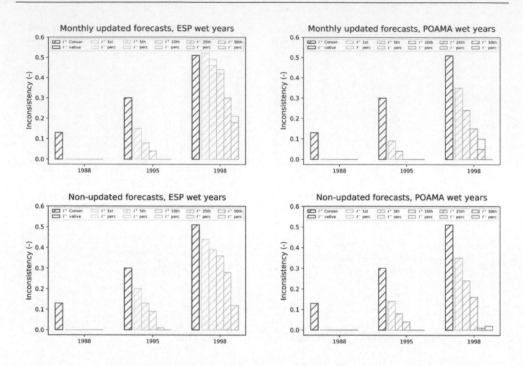

Figure 5.13. Inconsistency for selected wet years (1988, 1995 and 1998) using new inflow predictions every month (Monthly updated forecasts) and one inflow prediction for the next 12 months (Non-updated forecasts).

For those wet years (Figure 5.11) where the initial water estimate is below 100% (e.g. 1995 and 1998), there are primarily upward revisions of the allocation (with only very minor downward revisions) for all percentiles of the forecast ensemble. Water allocation results using the forecast ensemble are generally equal to the perfect water allocation after the September 14th decision date for all non-exceedance percentiles until the end of the cropping season. For several wet years, the water allocation decision based on the conservative inflow are lower than for the perfect water allocation at the beginning of the water year and then progressively increase (e.g. 1982 and 2003). These may be up to 55% lower than the perfect water allocation of 100% (e.g. 1998). For all years where this occurs, using the forecast ensemble shows better water allocation, even for the lowest forecast percentiles, which are closest to the conservative forecast. For many of the wet years, the initial allocation using both the perfect information, as well as the conservative inflow are already at 100%, as are allocations based on all forecast percentiles.

The annual inconsistency (number of upward revisions) using the forecast ensemble (see Figure 5.12 and Appendix B) is lower than when using the conservative inflow for the 10th, 25th and 50th non-exceedance percentiles. For the wet years, using the least conservative (50th percentile) leads to the lowest annual inconsistency (see Figure 5.13 and Appendix B). Although the annual inconsistency provides information about how the upward and downward revisions during the water year, it does not allow for easy comparison against the perfect water allocation. To evaluate how close the allocations established using the forecast ensemble were to those using the perfect information the Root Mean Squared Difference (RMSD) was calculated using difference between the water allocation established using the ensemble and that of the perfect forecast for each decision date. In Figure 5.14 and Figure 5.15, the RMSD for each decision date is shown for dry and wet years using the forecast ensemble, as well as for the allocations based on the conservative inflow.

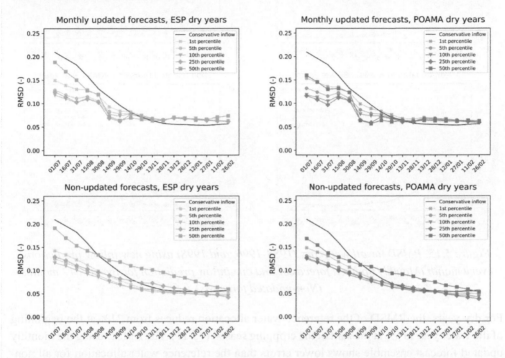

Figure 5.14. RMSD for all dry years (1982, 1997, and 1999-2009) using new inflow predictions every month (Monthly updated forecasts) and one inflow prediction for the next 12 months (Non-updated forecasts).

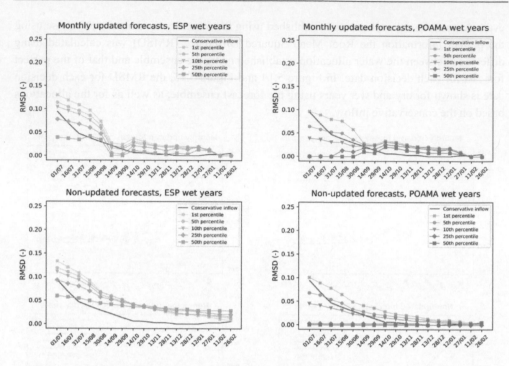

Figure 5.15. RMSD for all wet years (1983-1996, and 1998) using new inflow predictions every month (Monthly updated forecasts) and one inflow prediction for the next 12 months (Non-updated forecasts).

For dry years, the RMSD of the reference water allocation reduces from 21% at the beginning of the water year to 6% at the start of the cropping season (13/11). The RMSD using the monthly updated forecast ensemble shows lower errors than the reference water allocation for all non-exceedance percentiles until October 14th (ESP) and October 29th (POAMA). On October 14th (ESP and POAMA) using the 1st percentile of the ensemble results in a higher error compared to the reference water allocation. The RMSD using the monthly updated forecast ensemble is lower than the RMSD using the conservative inflow between July 1st and September 29th. A major error reduction occurs between August 30th and September 14th (from 13% to 6% for the 50th percentile POAMA). After September 14th the difference remains more or less equal until the end of the cropping season.

For wet years, the RMSD of the reference water allocation reduces from 9% at the beginning of the water year to below 5% in the cropping season (from 13/11 until 26/02). The RMSD in water allocation using the monthly updated forecast ensemble shows lower errors than the reference water allocation (using the conservative inflow), especially when using the 50th percentile. The RMSD in water allocation using the 50th percentile tends towards the RMSD of the reference water allocation after September 14th. For lower non-exceedance percentiles, the error increases. After September 14th the error is similar between non-exceedance percentiles.

5.4 DISCUSSION

The water allocation framework developed in this study was applied to assess the benefit of using a seasonal forecast ensemble (FoGSS) in estimating the available water for allocation. The potential for farmers to benefit from this seasonal forecast is discussed in the following paragraphs based on i) how well climate can be predicted, ii) how much this information helps in the actual decision process and iii) how much it contributes in reducing the negative impacts (Hansen, 2002).

5.4.1 How well can the climate be predicted?

FoGSS is an experimental seasonal streamflow forecast ensemble for a 12 months horizon developed for the Australian continent (Bennett et al., 2016, 2017) . FoGSS post-processes climate forecasts, either derived from ESP or POAMA to force a monthly hydrological model. ESP is an ensemble of seasonal precipitation forecasts based on climatology and POAMA is an ensemble including a coupled ocean-atmosphere general circulation model (CGCM).

A long time series (28 years) of climate predictions was used assuming a full representation of climate variability. The time period includes extremely dry years between 2001 and 2009 referred to as the Millennium Drought (van Dijk et al., 2013b) and very wet years, such as 1988 and 1989 (BoM, 2019b).

The derived inflow predictions were transformed into accumulative monthly volumes from each prediction month until the end of the water year (June). This set up was used to mimic the current water allocation process in which the basin authority uses a conservative inflow prediction at the beginning of the water year for the next 12 months. In our approach a prediction of the inflow to the end of the season is updated in each month using the forecast ensemble.

Our findings reveal that for the prediction months the inflow forecast ensemble contain a mixture of high and low biases, both for the FoGSS forecast based on ESP and on POAMA. We would expect progressively better results after the first prediction month in July, as the climate forecast is updated for a progressively shorter prediction period. However, this is not reflected in the inflow ensemble results. Uncertainty of the predicted inflows in July and August, at the start of the wet season are dominated by the uncertainty in the precipitation forecast from POAMA/ESP. In the sixth prediction month (December), the worst expected inflow is, however, obtained. We argue that this happens due to the memory of the hydrological model used to simulate basin conditions from previous wet months. October and November are wet months (BoM, 2019a), with a high variability for the selected years (1982-2009). This can explain the under-dispersion of the ensemble in December, as the seasonal forecast then contains little rainfall, with simulated flows conditioned primarily by (uncertain) model states at the start of the forecast, which are treated deterministically. Inflow predictions after February do progressively improve due to the shorter prediction periods, as the updated climate forecasts compensates the model uncertainty. Further evaluation is needed to determine how these biases negatively affect the water allocation decisions.

5.4.2 How much does FoGSS help in the actual decision process?

Currently, only upward water allocation revisions are made during the water year in river basins in Australia. This is because a conservative inflow for the next 12 months is used to determine the water allocation, essentially resulting an estimate of available water that is always on the safe side from the point of view of the allocation decision. However, for the farmer, advance knowledge of the ultimate allocation is beneficial as it allows for better planning. Using information from a seasonal forecast ensemble to provide a better estimate of inflows to reservoirs to inform the water allocation decision may reduce upward revisions and lead to improved allocation decisions. Our water allocation framework and findings using FoGSS add value to the typical forecast skill assessment. Particularly, the novelty of our work consists in arranging the forecasted datasets in the way that inflow predictions are used for the next n months until the end of the water year in a water allocation framework. In addition, key decision dates for water allocation are evaluated to determine the impact on farmers planning and operational decisions, especially for months before and during the cropping season.

We found that using the seasonal forecast ensemble does not reduces upward revisions (or positive inconsistency) for all years, but, that it does improve the accuracy in estimating available water for allocation (lower RMSD). Hence using the seasonal forecast ensemble helps in the actual decision process because the available water during the water year is better predicted. However, this comes at a cost. There is a trade-off between obtaining better predictions of water available for allocation than the conservative low estimate as it comes at the cost of more downward decisions during the water year. This is more evident for dry years than for wet years. In dry years downward revisions of the water allocation due to forecast biases. In wet years upward revisions are reduced (compared to the reference water allocation) and only very minor downward revisions occur. For both dry and wet years, the accuracy of the available water for allocation using the forecast ensemble improves during the water year. This is most evident on the decision date around September 14th, where the accuracy of the water allocation decision informed by the forecast improves significantly and maintains this accuracy until the end of the water year. The benefit of using the forecast ensemble is shown clearly in the root mean square difference results, which shows the magnitude of the difference in allocation when compared to that made with the perfect forecast. The root mean square difference with the forecast ensemble attains a value more or less to the root mean square difference obtained with the conservative inflow, but two months before the original decision date (from November 13th to September 14th). This means that for the same accuracy of predicted water allocation, decision makers can rely on the forecast ensemble information two months in advance when compared to the conservative inflow information. This may be of significant benefit to the farmers as they can then better plan their irrigation season based on the amount of water they would expect to be allocated.

Whether the basin authority in charge of water allocation announcements would choose to adopt a seasonal forecast such as that provided by FoGSS and change the water allocation policy will depend very much on how acceptable downward revisions of allocated water are. The current policy has been designed to avoid such downward revisions unless exceptional circumstances dictate. When using the full potential of the seasonal forecast (including a monthly update of the expected inflow as new forecasts become available) provides more accuracy in water allocation estimates at the cost of downward revisions. How acceptable such downward revisions may be will depend on the impact these have, compared to the benefit of the improved and more advanced information on the water allocation. The ratio of downward and upward revisions is also influenced by the selection of the non-exceedance percentile. For the dry years, selecting a lower, more conservative, percentage would appear to be the best strategy, while for the wet years a much higher percentage should be selected. A non-exceedance percentile of 10% would appear to provide the best performance on average for both dry and wet years evaluated in this study, but a more dynamic approach could also be taken depending on the forecast as well as the available storage at the beginning of the water year. The downward revisions that occur (primarily in the dry years) are primarily due to biases in the inflow forecast. When comparing the results of forecasts made using the POAMA dataset to those of the ESP dataset, the number of revisions (both upward and downward) in Figures 12 and 13 indicate that these are marginally smaller for the POAMA dataset. This would suggest that further improving the seasonal forecast would contribute to reducing undesirable downward revision. Additional improvement to the inflow predictions through reducing the uncertainty in the hydrological model of the basin will also contribute to reducing the bias of the inflow predictions and improved allocation results.

5.4.3 How much does FoGSS contribute in reducing negative impacts?

The impact in irrigated agriculture of uncertainty in the available water resources has been widely assessed in Australia considering climate variability and climate change scenarios (Kirby et al., 2013, 2014a, 2014b, 2015). Adaptation measures, reallocation strategies and policy reform are currently in discussion to prevent future impacts due to extreme events (Bark et al., 2014; van Dijk et al., 2013b). In this study we explore the possibility of using a seasonal forecast ensemble to secure the right amount of allocated water at the right time during the water year. The contribution of using a seasonal forecast ensemble to provide more accurate estimates of the water allocation earlier in the season contributes to reducing production losses in irrigated agriculture. More accurate and earlier estimate of the water allocation can enhance the decisions made at farm level, such as choosing which crop to plant and the area to be irrigated.

In the Murrumbidgee basin, farmers decide on the area to be cropped based on the water allocation announcement for General Security. The water year starts on July 1st, but the summer cropping season starts on November 1st and ends March 1st. In that sense, the period to decide on the area to be cropped is between July 1st and November 1st, and the period for operational decisions (e.g. irrigation schedule, weed management) is from November 1st to March 1st. Farmers can wait until the last allocation announcement before November 1st to decide on the area to be cropped, but due to pre-cropping planning activities and investments (e.g. buying seeds, maintenance of irrigation assets, or investing in agricultural equipment and machinery) they would prefer to take decisions earlier, and rely on water allocation announcements made on an earlier date. Based on the findings of our work, we would recommend famers to rely on the water allocation announcement made on September 14th, when using the currently available forecasts to inform the water allocation decisions. This will reduce negative impacts due to either over or underestimating the area to be cropped as the available water is better predicted.

5.5 CONCLUSION

We apply a water allocation framework to assess the benefit of using a seasonal forecast ensemble in water allocation decisions. This water allocation framework uses an estimate of the available water for the irrigation season that is based on the balance of the demand to the available water in the reservoirs in a basin and the expected inflows to those reservoirs from the decision date until the end of the water year. Our framework emulates the decisions made by basin authorities on the allocation of water to meet claims to water as defined in water concessions. Depending on availability, water may be allocated to fully meet these concessions or only to a set percentage. We apply the framework in the Murrumbidgee basin in Australia. In this basin, conservatively low estimates of the expected inflow based on climatology are currently used at the beginning of the water year to estimate the available water. As the water year progresses, water allocated to each concession may be revised if expected water availability improves. As the initial estimates are conservative, water allocations are mostly conservatively low, and consequently revised upwards. Although upward revisions of the allocation is beneficial to irrigators, advance and consistent information on the water allocated is important to them to help better plan their irrigation season.

Instead of the currently used conservative low estimates for inflow predictions we propose using inflow predictions from an ensemble seasonal streamflow forecast to inform water allocation decisions. Inflow predictions are obtained from the "Forecast Guided Stochastic Scenarios" (FoGSS), a seasonal ensemble streamflow forecasting system for Australia using either Ensemble Streamflow Prediction (ESP) or the POAMA M2.4 seasonal climate forecasting system as climate forcing. Of the two, predicting the inflows into reservoirs using the POAMA datasets were found to have better skill than using the ESP datasets. The seasonal forecast ensemble helps improve the decision process because the available water estimates during the water year are better predicted than the conservative forecast used as reference. However, the inconsistency in water allocation decisions during the water year is higher when using the seasonal forecast. This implies establishing a trade-off between obtaining better accuracy in available water estimates and the cost of more downward revisions during the water year. This is more evident for dry years than for wet years. In dry years more downward decisions occur than is currently the case. In wet years upward revisions are reduced (compared to the reference water allocation) and relatively no downward revisions occur. Both for dry or for wet years, the accuracy of the available water estimates using the forecast ensemble improves progressively during the water year, especially some one and a half months before the start of the cropping season in November. This additional time is important to the user, as it allows farmers to better plan the cropping season (November to February). Using the forecast ensemble thus contributes to reducing negative impacts in irrigated agriculture. The use of these datasets would help the basin authority in developing a better estimate of the available water for allocation and reduce agricultural losses.

5. The benefit of using an ensemble of seasonal streamflow forecasts

6

SYNTHESIS AND RECOMMENDATIONS FOR FUTURE RESEARCH

6.1 SYNTHESIS

New hydrological datasets from earth observations, hydrological and land-surface models, as well as seasonal forecasts continue to be developed and tested around the world. The quality of these datasets has been improving rapidly due to the consideration of longer periods of record, increasingly finer spatial resolutions, better accuracy in hydrological estimates and predictions, and the consideration of uncertainty through ensembles (Derin et al., 2016; Dijk et al., 2016; Her et al., 2019; Orth et al., 2017; Schellekens et al., 2017). For example, a new version of MSWEP global precipitation dataset (historic dataset over 30 years) was enhanced from a 0.25 degree to a 0.1 degree spatial resolution in less than one year (Beck et al., 2017b, 2018). These improvements increase the potential application of these datasets in smaller basins to support estimations of surface water availability and the development of basin management plans. In addition, seasonal streamflow forecasts are developed and applied to support operational decisions. For example, the ECMWF seasonal forecasting system is used with improved skill from sea surface temperature (SST) prediction and the fifth generation of atmospheric reanalyses of the global climate (Copernicus Climate Change Service (C3S), 2017; Stockdale et al., 2011). In the Australian continent an experimental seasonal streamflow forecast ensemble has been developed to better predict streamflow for a 12 month horizon (Bennett et al., 2017). At the same time, these improved hydrological datasets have become more accessible through online portals and databases. End users such as basin and water authorities can easily download hydrological datasets and make use of them to support their decision-making processes. However, despite these improvements it is not fully clear what the value is of using these datasets for decision making. This thesis explores the value of using hydrological datasets to support decisions in managing water resources, in particular those related to water allocation. The limitations and potential of these datasets is addressed in view of various aspects. On the one hand, the potential value of these datasets depends on the current network of hydrological stations being used. If the there is a well established network, with a good density of hydrological measurements the value of additional datasets may well be limitied. On the other hand, the actual value of hydrological datasets depends on the scale and purpose of the decision. If, for example, the decision maker would like to estimate the adequate design area of an irrigation district based on the available water to satisfy irrigation requirements, a monthly time scale would be enough for the assessment. However, if the decision maker would like to know about the available water for weekly water allocation announcements, a daily times scale would be necessary for an adequate evaluation.

6.1.1 Hydrological data quality and water allocation decisions

The first step was to evaluate the existing network of hydrological stations being used in large irrigation districts. In chapter 2, the link between sub-optimal water allocation decisions and the quality and availability of hydrological information in irrigation districts is explored. In this case, the water allocation decision that is considered is an operational decision in an irrigation district which consists of the opening or closing the main canal gates at the water source location to meet downstream demand, while considering the availability of water. An example of a sub-optimal water allocation decision is where there is sufficient water available to satisfy irrigation demand, but the operator does not open the gate on the perception that there is insufficient water, thus generating water shortage in the canal system. It was found that such sub-optimal water allocation decisions are often taken in large irrigation districts in Australia, Costa Rica and Colombia. The rate of which which sub-optimal water allocation decisions are taken is found to depend on the quality of the hydrological information available. In an irrigation district with high quality hydrological information, the rate of occurrence of sub-optimal water allocation decisions is relatively low when compared to an irrigation district with low quality hydrological information. The quality of hydrological information is quantified by an index composed of the period of record, spatial and temporal resolution of the available in-situ data at different locations in the irrigation district and the water supply basin. The challenge is to reduce sub-optimal water allocation decisions through complementing the in-situ data available with additional hydrological information that can help extend the period of record, and improve the temporal and spatial resolution. The key message is to explore the use of hydrological information additional to the available in-situ measurements, including hydrological models, remote sensing and reanalysis datasets, and through this help reduce sub-optimal water allocation decisions.

6.1.2 The value of using additional hydrological data from global observations and models

Global precipitation datasets and global hydrological models are continuously being developed, tested and improved. In chapter 3, the value of using an ensemble of global hydrological models is evaluated in determining the surface water availability for irrigation area planning. In chapter 4, the value of two global precipitation datasets, CHIRPS and MSWEP, is assessed for the same purpose, but now in combination with calibrated local hydrological models.

A hydro-economic framework was developed and applied to evaluate the value of using longer periods of record as well as of ensembles of river discharge information from global precipitation datasets and global hydrological models in irrigated agriculture. This framework combines principles of the economic utility theory (Neumann and Morgenstern, 1966) and probabilitstic hydrology (Wilks, 2011) applied to irrigation management. The Relative Utility Value is introduced as a measure to determine the risk of annual crop production loss based on the monthly probability of water scarcity for the derived size (area) of an irrigation district. The size of these irrigation areas was derived using the simulated river discharge information and compared against a reference area that was established using the available observed river discharge information. The annual crop yield loss was evaluated for rice, which is the main crop grown in the case studies explored, when water scarcity occurs in each growth stage. The Relative Utility Value varies depending on the monthly probability of water scarcity, but also on the monthly crop sensitivity due to water deficit.

Initially, the hydro-economic framework was tested in a large basin in Australia (84000 km²) using information from global hydrological models. It was found that using an ensemble of global hydrological models is more beneficial than using single models because the information

117

content of the ensemble provides a more robust estimate of the area that can be reliable irrigated, given available water. This means that a shorter time period of river discharge information can be used when using an ensemble of global hydrological models as the information content of the ensemble offsets the reduction of the time period. Specifically, using an ensemble of global hydrological models with a period of fifteen years of river discharge information provides equal accuracy in irrigated area estimation compared to thiry years of river discharge information from single global hydrological models.

The hydro-economic framework was extended to explore the cost of choosing between irrigation areas obtained from simulated river discharge calibrated with short time periods of observed river discharge. The simulated river discharge is determined by using different global precipitation products (e.g. CHIRPS and MSWEP) for a reference thirty year period driven by calibrated hydrological models. The rationale of the approach is that a modeller would like to know what happens if only five years of river discharge are available to calibrate a model. To emulate this situation, six independent samples, each five years in length, were extracted from the reference thirty-year dataset for calibration of the model parameters. Prior information about which period of five years is available for calibration and how representative each five-year sample is of the hydro-climatic variability was assumed to be unknown. Among the six periods extracted from the full thirty year dataset, some are sampled from more wet periods, while others represent normal or dry periods. Once a period of five years is selected the calibration is done, resulting in a model with the "best" model parameter sets, conditional on the sample used for calibration. Simulated river discharges are then obtained for the reference thirty year period using global precipitation datasets. The approach was tested in a medium sized (2000 km²) mountainous basin in Colombia. The main outcome is that it is more cost-effective to plan irrigated areas using thiry years of simulated river discharge information from global precipitation datasets than using only five years observed discharge data. This happens because the additional precipitation data from five to thirty years constrains the model uncertainty of the calibration procedure. In the selected basin, using additional precipitation data from CHIRPS outperforms MSWEP. Both precipitation datasets are originally corrected with information from ground stations, but in the selected basin the amount of stations used for correction is lower for MSWEP. In addition, MSWEP has a coarser resolution than CHIRPS, contributing to lower performance.

The application of an ensemble of global hydrological models allows the use of shorter time periods in river discharge estimates compared to single global hydrological models. On the other hand, the application of global precipitation datasets helps in providing longer time periods allowing the simulation of reliable surface water estimates when observed river discharge is limited. In large basins, the application of an ensemble of global hydrological models is recommended for initial surface water estimations for irrigation area planning. In corresponding subbasins, the application of a calibrated hydrological model is suitable to estimate the area to be cropped in irrigated agriculture.

6.1.3 The value of using seasonal streamflow forecasts

In chapter 3 and chapter 4 the value of using different global hydrological datasets were evaluated to enhance planning decisions in irrigated agriculture. These decisions include long term planning decisions such as the viability of developing new irrigation areas. However, there is also added value to using hydrological datasets to enhance operational decisions. In chapter 5, the value of using seasonal streamflow forecasts is assessed. The use of seasonal forecasts to support decisions has been addressed in several settings (Crochemore et al., 2016; Shukla et al., 2014; Turner et al., 2017; Winsemius et al., 2014). However, the potential enhancement in water allocation decisions in irrigated agriculture, informed by seasonal forecasts has been little studied. A complete assessment of the added value of seasonal forecasts can allow basin authorities to explore the opportunities seasonal forecasts provide to improve their operational decisions, and reduce potential losses to agricultural by improving water allocation estimates. In many basins, conservative estimates of the available resource in rivers and reservoirs may lead to allocated water to be underestimated due to conservative estimates of future (in) flows. Water allocations may be revised as the season progresses. The resulting inconsistency in allocation is undesirable to farmers and may lead to opportunity costs in agricultural production. In chapter 5, a feedback loop between simulated reservoir storage and decisions on the amount of water to allocate to irrigation was developed to evaluate two seasonal forecast datasets (Poama and ESP) derived from the Forecast Guided Stochastic Scenarios (FoGSS), a 12 month seasonal ensemble forecast in Australia (Bennett et al., 2017). The approach was evaluated in the Murrumbidgee basin, comparing water allocations obtained with an expected reservoir inflow from FoGSS against the allocations obtained with an expected reservoir inflow from a conservative climatological estimate (as currently used by the basin authority), as well as against those obtained using observed inflows (perfect information). The inconsistency in allocated water is evaluated the changes in allocated water made every 15 days from the initial allocation at the start of the water year to the end of the irrigation season, including both downward and upward revisions of allocations. Results show that the inconsistency due to upward revisions in allocated water is lower when using the forecast datasets (Poama and ESP) compared to the conservative inflow estimates (reference) which is beneficial to the planning of cropping areas by farmers. Over confidence can, however, lead to an increase in undesirable downward revisions. Even though biases are found in inflow predictions, the accuracy of the available water estimates using the forecast ensemble improves progressively during the water year, especially one and a half months before the start of the cropping season in November. This means that basin authorities and farmers can rely on the forecast ensemble for water allocation decisions, as these enhance the accuracy of the available water estimates, and importantly, provide more time to prepare for pre-cropping activities and investments (e.g. buying seeds, maintenance of irrigation assets, or investing in agricultural equipment and machinery) and plan adequate cropping areas.

6.2 RECOMMENDATIONS FOR FUTURE RESEARCH

Although the research shows the significant potential of using hydrological modelling and earth observations combined with in-situ datasets, and application of a seasonal forecast the framework developed in this study contains several limitations that should be further addressed such as the uncertainty in hydrological models, and the conjunctive water use between surface water and groundwater in irrigation districts.

In this research, it was found that an ensemble of global hydrological models is beneficial in estimating the surface water availability in a large basin in Australia. In addition, a local calibrated hydrological model showed to be useful in estimating the surface water availability in a medium sized basin in Colombia. However, the hydrological models are under continuous process of improvement. Global hydrological models, such as LISFLOOD or PCR-GLOBWB, lack information about hydropower operation and storage change. Local hydrological models, such as the one used in Colombia (Dynamic Water Balance Model), are based only on three model parameters, which may underestimate the full characterization of the basin. In the Magadelana-Cauca basin in Colombia, three hydrological models including the Dynamic Water Balance Model, VIC and WFLOW models were tested in several subbasins for water resources management purposes (Rodriguez et al., 2017). These results may help in deciding which hydrological model to use for a specific subbasin in Colombia.

In the evaluated irrigation districts the conjunctive use between surface water and groundwater was not considered. It is crucial to continue work and efforts to understand the groundwater use in these irrigation districts and how it is affecting the water balance and groundwater levels. Several studies suggests that groundwater is not being fully monitored and the available resource has not been properly quantified (Cuadrado-Quesada et al., 2018; Petheram et al., 2008). In Costa Rica, quantifying the total available resource (including surface and groundwater) may support water allocation decisions and avoid unnecessary investments.

In addition, the framework used to estimate the crop yield due to water scarcity can be extended in order to include more characteristics such as irrigation management practices, and the actual crop stage development. Irrigation management practices such as transplanting or direct sowing, fertilizers use or even different crop varieties may affect crop yield. The actual crop stage development will help in estimating the true evapotranspiration deficit. The application of an open source crop model such as AquaCrop from FAO (Raes et al., 2009; Steduto et al., 2009) and AquaCrop-OS (Foster et al., 2017) can be explored.

Finally, the collaboration between different fields of study is recommended. An effort was done in this PhD work to combine knowledge about water allocation policy and technical information about hydrological datasets. However, similar efforts should be developed in larger countries and at basin level and irrigation district level in order to gain accountability about the use of hydrological data. Researchers and private consultancy may know that the use of hydrological data is important, however if current policy and regulation do not encourage and set requirements on the proper spatial and temporal resolution of the data, possibly the estimation of the water resource will be established based on inaccurate and low quality information, which will finally impede the adequate planning of the available resource and lead to water conflicts.

6. Synthesis and recommendations for future research

APPENDIX A

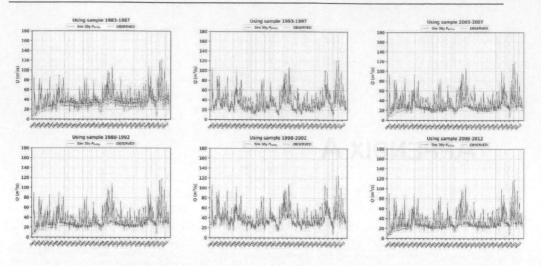

Figure A1. Observed and simulated Coello River discharge with 30 years (1983-2012) of In-Situ precipitation (Sim 30y $P_{In-Situ}$) for all calibration samples (1983-1987, 1988-1992, 1993-1997, 1998-2002, 2003-2007, 2008-2012).

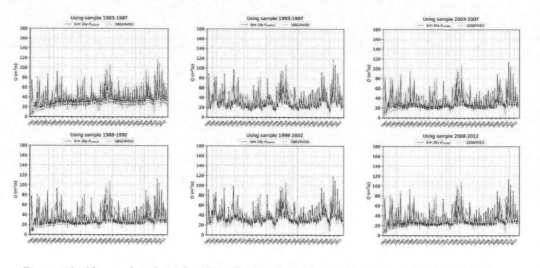

Figure A2. Observed and simulated Coello River discharge with 30 years (1983-2012) of CHIRPS precipitation (Sim 30y P_{CHIRPS}) for all calibration samples (1983-1987, 1988-1992, 1993-1997, 1998-2002, 2003-2007, 2008-2012).

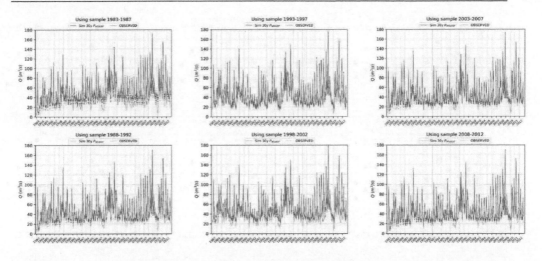

Figure A3. Observed and simulated Coello River discharge with 30 years (1983-2012) of MSWEP precipitation (Sim 30y P_{MSWEP}) for all calibration samples (1983-1987, 1988-1992, 1993-1997, 1998-2002, 2003-2007, 2008-2012).

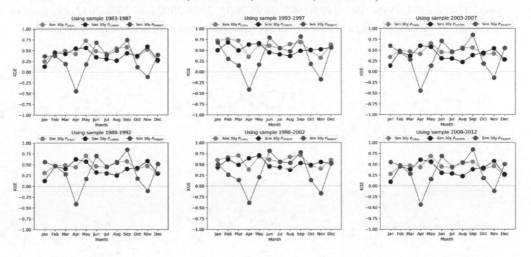

Figure A4. KGE performance metric for simulated river discharge for the complete time period of 30 years (1983-2012) using three different precipitation datasets (In-Situ, CHIRPS and MSWEP) in the Coello basin. Six independent samples of observed river discharge of 5 years (1983-1987, 1988-1992, 1993-1997, 1998-2002, 2003-2007 and 2008-2012) are used to calibrate model parameters, with the sample used for model calibration indicated in the header.

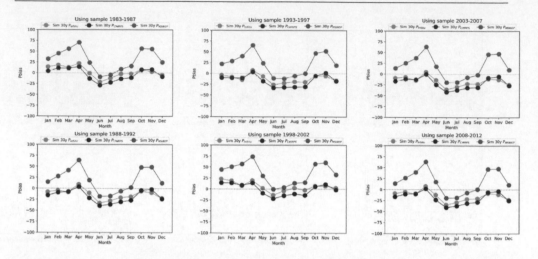

Figure A5. Pbias performance metric for simulated river discharge for the complete time period of 30 years (1983-2012) using three different precipitation datasets (In-Situ, CHIRPS and MSWEP) in the Coello basin. Six independent samples of observed river discharge of 5 years (1983-1987, 1988-1992, 1993-1997, 1998-2002, 2003-2007 and 2008-2012) are used to calibrate model parameters, with the sample used for model calibration indicated in the header.

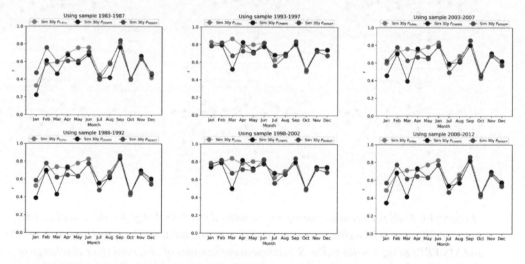

Figure A6. Pearson correlation, r performance metric for simulated river discharge for the complete time period of 30 years (1983-2012) using three different precipitation datasets (In-Situ, CHIRPS and MSWEP) in the Coello basin. Six independent samples of observed river discharge of 5 years (1983-1987, 1988-1992, 1993-1997, 1998-2002, 2003-2007 and 2008-2012) are used to calibrate model parameters, with the sample used for model calibration indicated in the header.

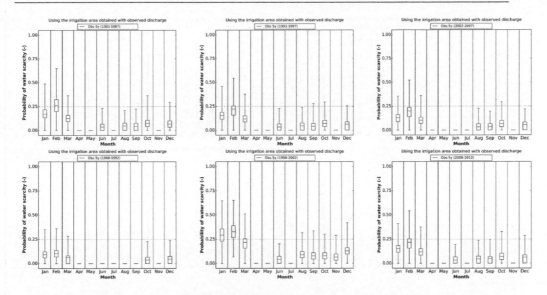

Figure A7. Probability of water scarcity using the irrigation area obtained with the observed river discharge of 5 years (Obs 5y) and the reference surface water availability. Boxplots show the median, interquartile range and minimum-maximum range.

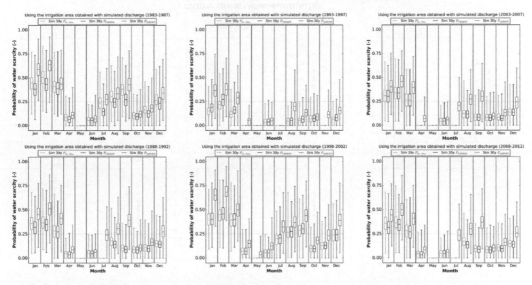

Figure A8. Probability of water scarcity using the irrigation area obtained with simulated river discharge information (Sim 30y $P_{In-Situ}$, Sim 30y P_{CHIRPS}, Sim 30y P_{MSWEP}) and the reference surface water availability. Boxplots show the median, interquartile range and minimum-maximum range.

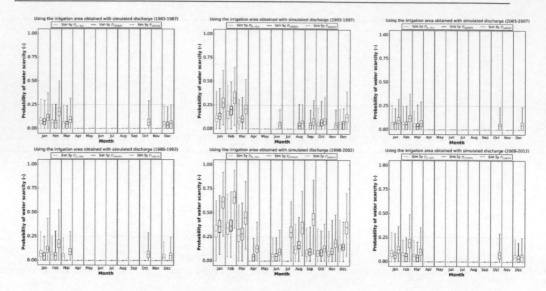

Figure A9. Probability of water scarcity using the irrigation area obtained with simulated river discharge information (Sim 5y $P_{In-Situ}$, Sim 5y P_{CHIRPS}, Sim 5y P_{MSWEP}) and the reference surface water availability. Boxplots show the median, interquartile range and minimum-maximum range.

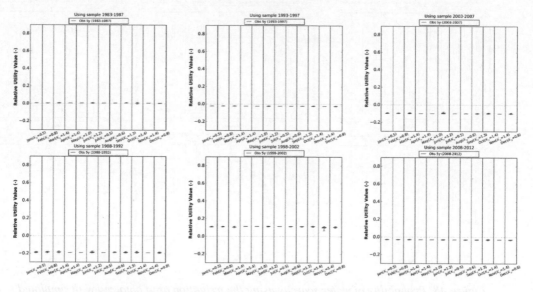

Figure A10. Relative Utility Value using observed river discharge of 5 years for water scarcity happening independently in one month. Ky is the sensitivity of the crop to water deficit. Boxplots show the median, interquartile range and minimum-maximum range.

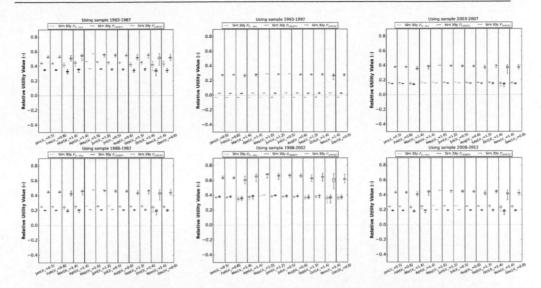

Figure A11. Relative Utility Value using simulated river discharge of 30 years for water scarcity happening independently in one month. Ky is the sensitivity of the crop to water deficit. Boxplots show the median, interquartile range and minimum-maximum range.

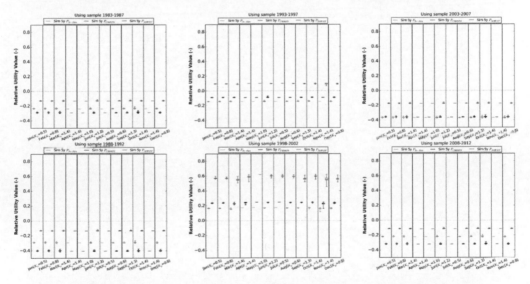

Figure A12. Relative Utility Value using simulated river discharge of 5 years for water scarcity happening independently in one month. Ky is the sensitivity of the crop to water deficit. Boxplots show the median, interquartile range and minimum-maximum range.

APPENDIX B

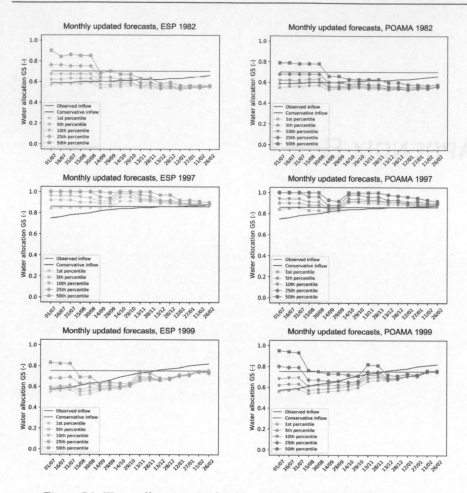

Figure B1. Water allocation GS for dry years (1982, 1997, 1999) using new inflow predictions every month (Monthly updated forecasts).

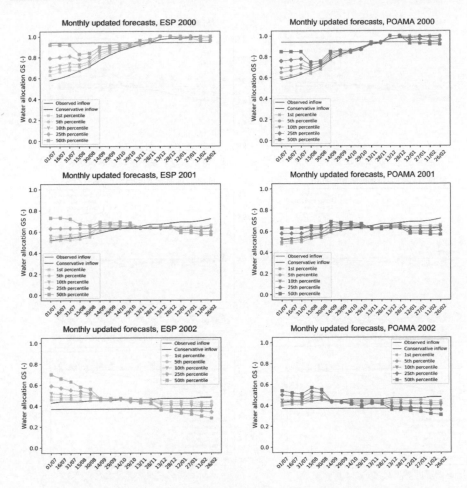

Figure B2. Water allocation GS for dry years (2000, 2001, 2002) using new inflow predictions every month (Monthly updated forecasts).

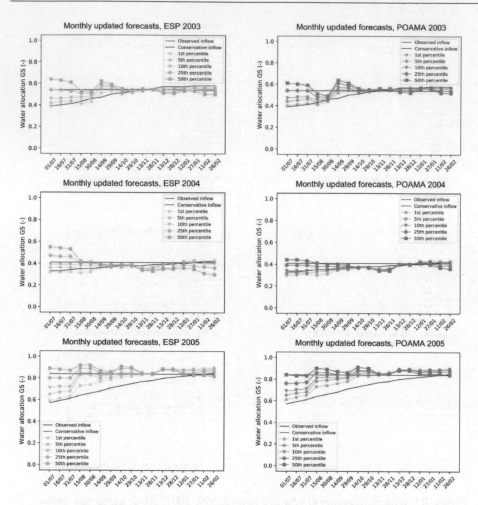

Figure B3. Water allocation GS for dry years (2003, 2004, 2005) using new inflow predictions every month (Monthly updated forecasts).

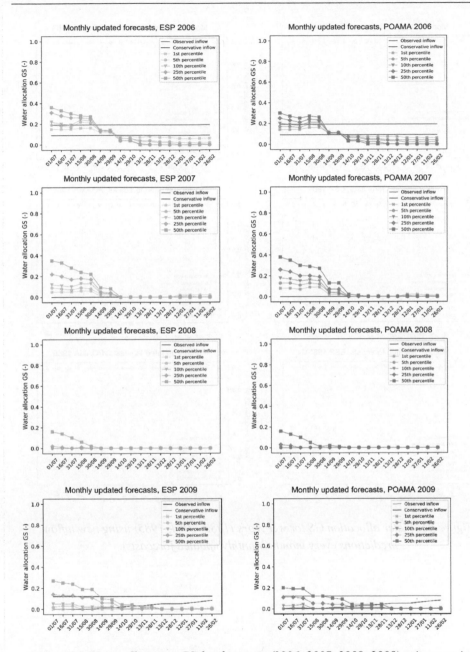

Figure B4. Water allocation GS for dry years (2006, 2007, 2008, 2009) using new inflow predictions every month (Monthly updated forecasts).

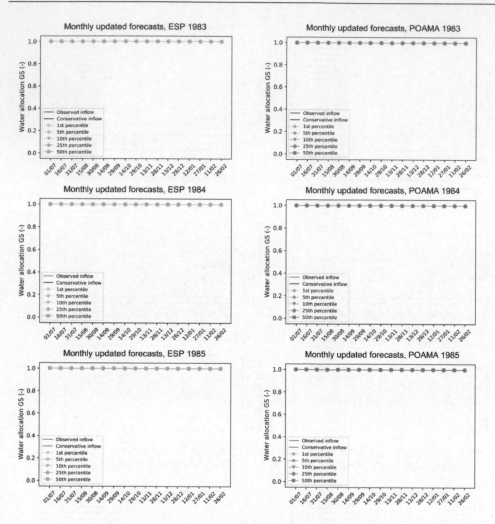

Figure B5. Water allocation GS for wet years (1983, 1984, 1985) using new inflow predictions every month (Monthly updated forecasts).

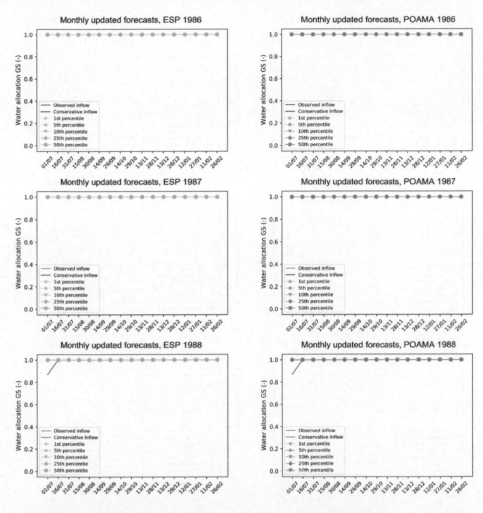

Figure B6. Water allocation GS for wet years (1986, 1987, 1988) using new inflow predictions every month (Monthly updated forecasts).

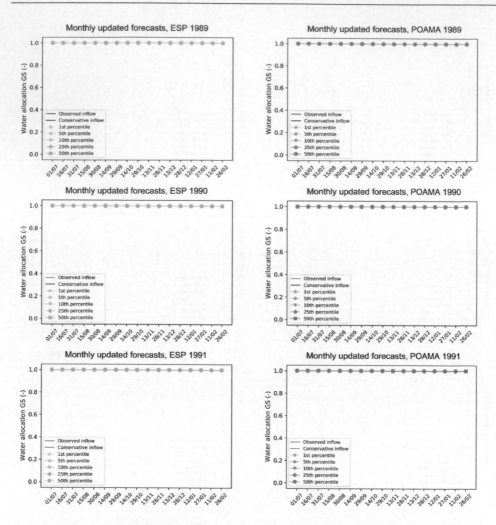

Figure B7. Water allocation GS for wet years (1989, 1990, 1991) using new inflow predictions every month (Monthly updated forecasts).

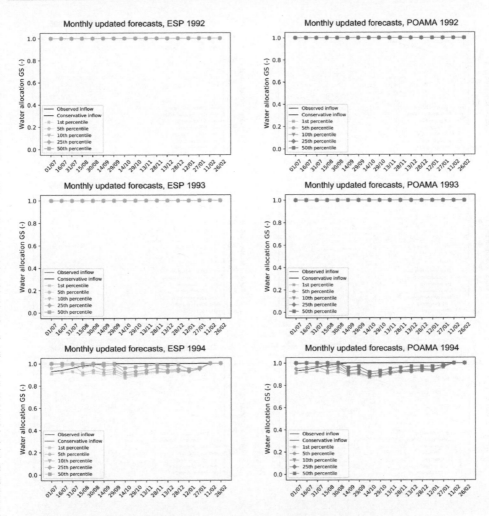

Figure B8. Water allocation GS for wet years (1992, 1993, 1994) using new inflow predictions every month (Monthly updated forecasts).

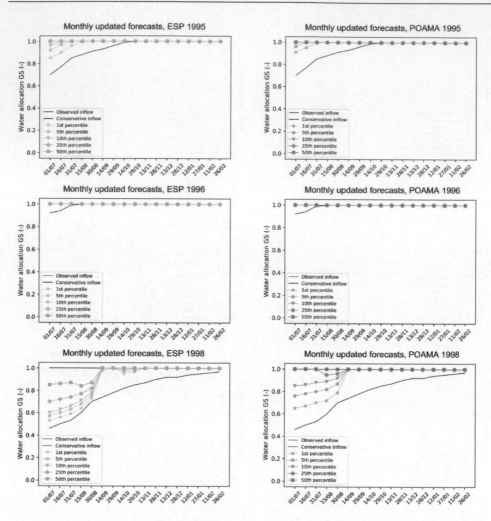

Figure B9. Water allocation GS for wet years (1995, 1996, 1998) using new inflow predictions every month (Monthly updated forecasts).

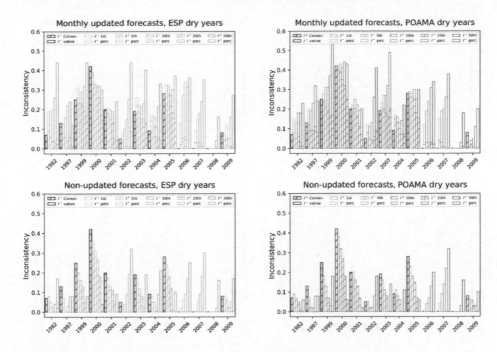

Figure B10. Inconsistency for dry years using new inflow predictions every month (Monthly updated forecasts) and one inflow prediction for the next 12 months (Non-updated forecasts).

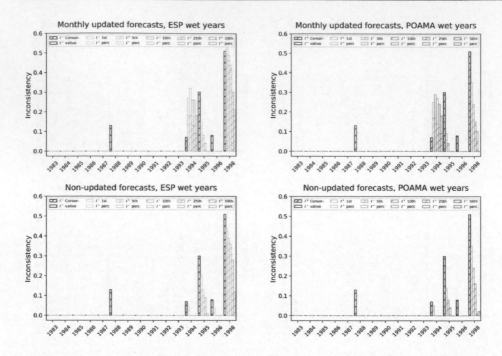

Figure B11. Inconsistency for wet years using new inflow predictions every month (Monthly updated forecasts) and one inflow prediction for the next 12 months (Non-updated forecasts).

REFERENCES

Al-Faraj, F. A. M. and Scholz, M.: Assessment of temporal hydrologic anomalies coupled with drought impact for a transboundary river flow regime: The Diyala watershed case study, J. Hydrol., 517, 64–73, doi:10.1016/j.jhydrol.2014.05.021, 2014.

Alfonso, L. and Price, R.: Coupling hydrodynamic models and value of information for designing stage monitoring networks, Water Resour. Res., 48(8), W08530, doi:10.1029/2012WR012040, 2012.

Anghileri, D., Voisin, N., Castelletti, A., Pianosi, F., Nijssen, B. and Lettenmaier, D. P.: Value of long-term streamflow forecasts to reservoir operations for water supply in snow-dominated river catchments, Water Resour. Res., 52(6), 4209–4225, doi:10.1002/2015WR017864, 2016.

Australian government: Attachment A - Murray-Darling Basin Reform - Memorandum of Understanding | Council of Australian Governments (COAG), [online] Available from: https://www.coag.gov.au/sites/default/files/agreements/Attachment-A-Murray-Darling-Basin-Reform-MOU.pdf (Accessed 20 August 2018), 2008.

Balsamo, G., Beljaars, A., Scipal, K., Viterbo, P., van den Hurk, B., Hirschi, M. and Betts, A. K.: A Revised Hydrology for the ECMWF Model: Verification from Field Site to Terrestrial Water Storage and Impact in the Integrated Forecast System, J. Hydrometeorol., 10(3), 623–643, doi:10.1175/2008JHM1068.1, 2009.

Bark, R., Kirby, M., Connor, J. D. and Crossman, N. D.: Water allocation reform to meet environmental uses while sustaining irrigation: a case study of the Murray–Darling Basin, Australia, Water Policy, 16(4), 739–754, doi:10.2166/wp.2014.128, 2014.

Baumüller, H.: The Little We Know: An Exploratory Literature Review on the Utility of Mobile Phone-Enabled Services for Smallholder Farmers, J. Int. Dev., 30(1), 134–154, doi:10.1002/jid.3314, 2018.

Beck, H. E., Vergopolan, N., Pan, M., Levizzani, V., van Dijk, A. I. J. M., Weedon, G. P., Brocca, L., Pappenberger, F., Huffman, G. J. and Wood, E. F.: Global-scale evaluation of 22 precipitation datasets using gauge observations and hydrological modeling, Hydrol Earth Syst Sci, 21(12), 6201–6217, doi:10.5194/hess-21-6201-2017, 2017a.

Beck, H. E., van Dijk, A. I. J. M., Levizzani, V., Schellekens, J., Miralles, D. G., Martens, B. and de Roo, A.: MSWEP: 3-hourly 0.25° global gridded precipitation (1979–2015) by merging gauge, satellite, and reanalysis data, Hydrol Earth Syst Sci, 21(1), 589–615, doi:10.5194/hess-21-589-2017, 2017b.

Beck, H. E., Wood, E. F., Pan, M., Fisher, C. K., Miralles, D. G., van Dijk, A. I. J. M., McVicar, T. R. and Adler, R. F.: MSWEP V2 global 3-hourly 0.1° precipitation: methodology and quantitative assessment, Bull. Am. Meteorol. Soc., doi:10.1175/BAMS-D-17-0138.1, 2018.

van Beek, L. P. H., Wada, Y. and Bierkens, M. F. P.: Global monthly water stress: 1. Water balance and water availability, Water Resour. Res., 47(7), W07517, doi:10.1029/2010WR009791, 2011.

Ben-Gal, A., Weikard, H.-P., Shah, S. H. H. and van der Zee, S. E. A. T. M.: A coupled agronomic-economic model to consider allocation of brackish irrigation water, Water Resour. Res., 49(5), 2861–2871, doi:10.1002/wrcr.20258, 2013.

Bennett, J., Wang Q. J., Li Ming, Robertson David E. and Schepen Andrew: Reliable long-range ensemble streamflow forecasts: Combining calibrated climate forecasts with a conceptual runoff model and a staged error model, Water Resour. Res., 52(10), 8238–8259, doi:10.1002/2016WR019193, 2016.

Bennett, J. C., Wang, Q. J., Robertson, D. E., Schepen, A., Li, M. and Michael, K.: Assessment of an ensemble seasonal streamflow forecasting system for Australia, Hydrol Earth Syst Sci, 21(12), 6007–6030, doi:10.5194/hess-21-6007-2017, 2017.

Best, M. J., Pryor, M., Clark, D. B., Rooney, G. G., Essery, R. L. H., Ménard, C. B., Edwards, J. M., Hendry, M. A., Porson, A., Gedney, N., Mercado, L. M., Sitch, S., Blyth, E., Boucher, O., Cox, P. M., Grimmond, C. S. B. and Harding, R. J.: The Joint UK Land Environment Simulator (JULES), model description – Part 1: Energy and water fluxes, Geosci Model Dev, 4(3), 677–699, doi:10.5194/gmd-4-677-2011, 2011.

Bierkens, M. F. P.: Global hydrology 2015: State, trends, and directions, Water Resour. Res., 51(7), 4923–4947, doi:10.1002/2015WR017173, 2015.

Bloschl, G., Sivapalan, M. and Wagener, T.: Runoff Prediction in Ungauged Basins: Synthesis Across Processes, Places and Scales, Cambridge University Press, Cambridge., 2014.

BoM: Climate statistics for Australian locations, [online] Available from: http://www.bom.gov.au/jsp/ncc/cdio/cvg/av?p_stn_num=070014&p_prim_element_index=18&p_display_type=statGraph&period_of_avg=ALL&normals_years=allYearOfData&staticPage=, 2019a.

BoM: La Niña – Detailed Australian Analysis, [online] Available from: http://www.bom.gov.au/climate/enso/lnlist/, 2019b.

Boucher, M.-A., Tremblay, D., Delorme, L., Perreault, L. and Anctil, F.: Hydro-economic assessment of hydrological forecasting systems, J. Hydrol., 416–417, 133–144, doi:10.1016/j.jhydrol.2011.11.042, 2012.

Bouma, J. A., van der Woerd, H. J. and Kuik, O. J.: Assessing the value of information for water quality management in the North Sea, J. Environ. Manage., 90(2), 1280–1288, doi:10.1016/j.jenvman.2008.07.016, 2009.

Budyko, M.: Climate and life, Academic Press, INC., New York, 1974.

Cerdá, E. and Quiroga, S.: Economic value of weather forecasting: the role of risk aversion, TOP, 19(1), 130–149, doi:10.1007/s11750-009-0114-3, 2011.

Chacon-Hurtado, J. C., Alfonso, L. and Solomatine, D. P.: Rainfall and streamflow sensor network design: a review of applications, classification, and a proposed framework, Hydrol Earth Syst Sci, 21(6), 3071–3091, doi:10.5194/hess-21-3071-2017, 2017.

Chang, F.-J. and Wang, K.-W.: A systematical water allocation scheme for drought mitigation, J. Hydrol., 507, 124–133, doi:10.1016/j.jhydrol.2013.10.027, 2013.

Chuchra, J.: Drones and Robots: Revolutionizing the Future of Agriculture, Geospatial World [online] Available from: https://www.geospatialworld.net/article/drones-and-robots-future-agriculture/ (Accessed 4 March 2019), 2016.

CONARROZ: Informe estadístico período 2013-2014, [online] Available from: https://www.conarroz.com/UserFiles/File/INFORME_ANUAL_ESTADISTICO_2014-2015.pdf, 2015.

Connor, J. D., Kandulu, J. M. and Bark, R. H.: Irrigation revenue loss in Murray–Darling Basin drought: An econometric assessment, Agric. Water Manag., 145, 163–170, doi:10.1016/j.agwat.2014.05.003, 2014.

Copernicus Climate Change Service (C3S): ERA5: Fifth generation of ECMWF atmospheric reanalyses of the global climate, 2017.

Córdoba-Machado, S., Palomino-Lemus, R., Gámiz-Fortis, S. R., Castro-Díez, Y. and Esteban-Parra, M. J.: Assessing the impact of El Niño Modoki on seasonal precipitation in Colombia, Glob. Planet. Change, 124, 41–61, doi:10.1016/j.gloplacha.2014.11.003, 2015.

Crochemore, L., Ramos, M.-H. and Pappenberger, F.: Bias correcting precipitation forecasts to improve the skill of seasonal streamflow forecasts, Hydrol Earth Syst Sci, 20(9), 3601–3618, doi:10.5194/hess-20-3601-2016, 2016.

Cuadrado-Quesada, G., Holley, C. and Gupta, J.: Groundwater Governance in the Anthropocene: A Close Look at Costa Rica, SSRN Scholarly Paper, Social Science Research Network, Rochester, NY. [online] Available from: https://papers.ssrn.com/abstract=3158875 (Accessed 15 August 2018), 2018.

Dai, Z. Y. and Li, Y. P.: A multistage irrigation water allocation model for agricultural land-use planning under uncertainty, Agric. Water Manag., 129, 69–79, doi:10.1016/j.agwat.2013.07.013, 2013.

DANE: 4° Censo Nacional Arrocero 2016, [online] Available from: https://www.dane.gov.co/index.php/estadisticas-por-tema/agropecuario/censo-nacional-arrocero (Accessed 15 June 2017), 2016.

Day, G. N.: Extended Streamflow Forecasting Using NWSRFS, J. Water Resour. Plan. Manag., 111(2), 157–170, doi:10.1061/(ASCE)0733-9496(1985)111:2(157), 1985.

Decharme, B., Alkama, R., Douville, H., Becker, M. and Cazenave, A.: Global Evaluation of the ISBA-TRIP Continental Hydrological System. Part II: Uncertainties in River Routing Simulation Related to Flow Velocity and Groundwater Storage, J. Hydrometeorol., 11(3), 601–617, doi:10.1175/2010JHM1212.1, 2010.

Decharme, B., Martin, E. and Faroux, S.: Reconciling soil thermal and hydrological lower boundary conditions in land surface models, J. Geophys. Res. Atmospheres, 118(14), 7819–7834, doi:10.1002/jgrd.50631, 2013.

Dee, D. P., Uppala, S. M., Simmons, A. J., Berrisford, P., Poli, P., Kobayashi, S., Andrae, U., Balmaseda, M. A., Balsamo, G., Bauer, P., Bechtold, P., Beljaars, A. C. M., van de Berg, L., Bidlot, J., Bormann, N., Delsol, C., Dragani, R., Fuentes, M., Geer, A. J., Haimberger, L., Healy, S. B., Hersbach, H., Hólm, E. V., Isaksen, L., Kållberg, P., Köhler, M., Matricardi, M., McNally, A. P., Monge-Sanz, B. M., Morcrette, J.-J., Park, B.-K., Peubey, C., de Rosnay, P., Tavolato, C., Thépaut, J.-N. and Vitart, F.: The ERA-Interim reanalysis: configuration and performance of the data assimilation system, Q. J. R. Meteorol. Soc., 137(656), 553–597, doi:10.1002/qj.828, 2011.

Derin, Y., Anagnostou, E., Berne, A., Borga, M., Boudevillain, B., Buytaert, W., Chang, C.-H., Delrieu, G., Hong, Y., Hsu, Y. C., Lavado-Casimiro, W., Manz, B., Moges, S., Nikolopoulos, E. I., Sahlu, D., Salerno, F., Rodríguez-Sánchez, J.-P., Vergara, H. J. and Yilmaz, K. K.: Multiregional Satellite Precipitation Products Evaluation over Complex Terrain, J. Hydrometeorol., 17(6), 1817–1836, doi:10.1175/JHM-D-15-0197.1, 2016.

Dijk, A. I. J. M. V., Brakenridge, G. R., Kettner, A. J., Beck, H. E., Groeve, T. D. and Schellekens, J.: River gauging at global scale using optical and passive microwave remote sensing, Water Resour. Res., 52(8), 6404–6418, doi:10.1002/2015WR018545, 2016.

van Dijk, A. I. J. M., Peña-Arancibia, J. L., Wood, E. F., Sheffield, J. and Beck, H. E.: Global analysis of seasonal streamflow predictability using an ensemble prediction system and observations from 6192 small catchments worldwide, Water Resour. Res., 49(5), 2729–2746, doi:10.1002/wrcr.20251, 2013a.

van Dijk, A. I. J. M., Beck, H. E., Crosbie, R. S., de Jeu, R. A. M., Liu, Y. Y., Podger, G. M., Timbal, B. and Viney, N. R.: The Millennium Drought in southeast Australia (2001–2009):

Natural and human causes and implications for water resources, ecosystems, economy, and society, Water Resour. Res., 49(2), 1040–1057, doi:10.1002/wrcr.20123, 2013b.

Döll, P., Fiedler, K. and Zhang, J.: Global-scale analysis of river flow alterations due to water withdrawals and reservoirs, Hydrol Earth Syst Sci, 13(12), 2413–2432, doi:10.5194/hess-13-2413-2009, 2009.

Dreverman, D.: Responding to Extreme Drought in the Murray-Darling Basin, Australia, in Drought in Arid and Semi-Arid Regions: A Multi-Disciplinary and Cross-Country Perspective, edited by K. Schwabe, J. Albiac, J. D. Connor, R. M. Hassan, and L. Meza González, pp. 425–435, Springer Netherlands, Dordrecht., 2013.

Eslamian, S.: Handbook of Engineering Hydrology: Modeling, Climate Change, and Variability, CRC Press [online] Available from: https://www.crcpress.com/Handbook-of-Engineering-Hydrology-Modeling-Climate-Change-and-Variability/Eslamian/p/book/9781466552463 (Accessed 25 October 2017), 2014.

FAO: Modernizing irrigation management - the MASSCOTE approach, [online] Available from: http://www.fao.org/docrep/010/a1114e/a1114e00.htm (Accessed 19 September 2014), 2007.

FAO: Climate change, water and food security, [online] Available from: http://www.fao.org/search/en/?cx=018170620143701104933%3Aqq82jsfba7w&q=climate+change+and+food+security&cof=FORID%3A9&siteurl=www.fao.org%2Fhome%2Fen%2F&ref=&ss=9087j2757481j46 (Accessed 2 February 2015), 2011.

FAO: Crop yield response to water, FAO Irrigation and Drainage Paper, Food and Agriculture Organization of the United Nations, Rome., 2012.

FAO, Ed.: Coping With Water Scarcity: An Action Framework For Agriculture And Food Security: FAO Water Reports No 38, Food & Agriculture Org., Rome., 2013.

FAO: Save and Grow in practice maize rice wheat - A Guide to Sustainable Cereal Production, 2016.

Far, S. T. and Rezaei-Moghaddam, K.: Impacts of the precision agricultural technologies in Iran: An analysis experts' perception & their determinants, Inf. Process. Agric., 5(1), 173–184, doi:10.1016/j.inpa.2017.09.001, 2018.

Fedearroz: Precio Promedio Mensual Arroz Paddy Verde en Colombia 2009-2016, [online] Available from: http://www.fedearroz.com.co/new/precios.php (Accessed 15 June 2017), 2017.

Flörke, M., Kynast, E., Bärlund, I., Eisner, S., Wimmer, F. and Alcamo, J.: Domestic and industrial water uses of the past 60 years as a mirror of socio-economic development: A global simulation study, Glob. Environ. Change, 23(1), 144–156, doi:10.1016/j.gloenvcha.2012.10.018, 2013.

Foster, T., Brozović, N., Butler, A. P., Neale, C. M. U., Raes, D., Steduto, P., Fereres, E. and Hsiao, T. C.: AquaCrop-OS: An open source version of FAO's crop water productivity model, Agric. Water Manag., 181, 18–22, doi:10.1016/j.agwat.2016.11.015, 2017.

de Fraiture, C. and Wichelns, D.: Satisfying future water demands for agriculture, Agric. Water Manag., 97(4), 502–511, doi:10.1016/j.agwat.2009.08.008, 2010.

Funk, C., Peterson, P., Landsfeld, M., Pedreros, D., Verdin, J., Shukla, S., Husak, G., Rowland, J., Harrison, L., Hoell, A. and Michaelsen, J.: The climate hazards infrared precipitation with stations—a new environmental record for monitoring extremes, Sci. Data, 2, sdata201566, doi:10.1038/sdata.2015.66, 2015.

Garces-Restrepo, C., Vermillion, D. and Muñoz, G.: Irrigation management transfer. Worldwide efforts and results, 2007.

Gerritsen, H.: What happened in 1953? The Big Flood in the Netherlands in retrospect, Philos. Trans. R. Soc. Lond. Math. Phys. Eng. Sci., 363(1831), 1271–1291, doi:10.1098/rsta.2005.1568, 2005.

Gosling, S. N., Zaherpour, J., Mount, N. J., Hattermann, F. F., Dankers, R., Arheimer, B., Breuer, L., Ding, J., Haddeland, I., Kumar, R., Kundu, D., Liu, J., Griensven, A. van, Veldkamp, T. I. E., Vetter, T., Wang, X. and Zhang, X.: A comparison of changes in river runoff from multiple global and catchment-scale hydrological models under global warming scenarios of 1 °C, 2 °C and 3 °C, Clim. Change, 141(3), 577–595, doi:10.1007/s10584-016-1773-3, 2017.

Green, D.: Water resources and management overview: Murrumbidgee catchment, Sydney, N.S.W.: NSW Office of Water. [online] Available from: http://trove.nla.gov.au/version/169998959 (Accessed 20 August 2018), 2011.

Gudmundsson, L., Tallaksen, L. M., Stahl, K., Clark, D. B., Dumont, E., Hagemann, S., Bertrand, N., Gerten, D., Heinke, J., Hanasaki, N., Voss, F. and Koirala, S.: Comparing Large-Scale Hydrological Model Simulations to Observed Runoff Percentiles in Europe, J. Hydrometeorol., 13(2), 604–620, doi:10.1175/JHM-D-11-083.1, 2011.

Gudmundsson, L., Wagener, T., Tallaksen, L. M. and Engeland, K.: Evaluation of nine large-scale hydrological models with respect to the seasonal runoff climatology in Europe, Water Resour. Res., 48(11), W11504, doi:10.1029/2011WR010911, 2012.

Gupta, H. V., Kling, H., Yilmaz, K. K. and Martinez, G. F.: Decomposition of the mean squared error and NSE performance criteria: Implications for improving hydrological modelling, J. Hydrol., 377(1–2), 80–91, doi:10.1016/j.jhydrol.2009.08.003, 2009.

Hanasaki, N., Fujimori, S., Yamamoto, T., Yoshikawa, S., Masaki, Y., Hijioka, Y., Kainuma, M., Kanamori, Y., Masui, T., Takahashi, K. and Kanae, S.: A global water scarcity assessment under Shared Socio-economic Pathways – Part 2: Water availability and scarcity, Hydrol. Earth Syst. Sci., 17(7), 2393–2413, doi:https://doi.org/10.5194/hess-17-2393-2013, 2013.

Hansen, J. W.: Realizing the potential benefits of climate prediction to agriculture: Issues, approaches, challenges, Agric. Syst., 74(3), 309–330, doi:10.1016/S0308-521X(02)00043-4, 2002.

Hargreaves, George H.: Defining and Using Reference Evapotranspiration, J. Irrig. Drain. Eng., 120(6), 1132–1139, doi:10.1061/(ASCE)0733-9437(1994)120:6(1132), 1994.

Harris, I., Jones, P. d., Osborn, T. j. and Lister, D. h.: Updated high-resolution grids of monthly climatic observations – the CRU TS3.10 Dataset, Int. J. Climatol., 34(3), 623–642, doi:10.1002/joc.3711, 2014.

Hellegers, P. and Leflaive, X.: Water allocation reform: what makes it so difficult?, Water Int., 40(2), 273–285, doi:10.1080/02508060.2015.1008266, 2015.

Her, Y., Yoo, S.-H., Cho, J., Hwang, S., Jeong, J. and Seong, C.: Uncertainty in hydrological analysis of climate change: multi-parameter vs. multi-GCM ensemble predictions, Sci. Rep., 9(1), 4974, doi:10.1038/s41598-019-41334-7, 2019.

Hirshleifer and Riley: The Analytics of Uncertainty and Information - An Expository Survey, [online] Available from: http://www2.uah.es/econ/MicroDoct/Hirs_Raley.pdf (Accessed 30 October 2014), 1979.

Hirshleifer, J. and Riley, J. G.: The Analytics of Uncertainty and Information-An Expository Survey, J. Econ. Lit., 17(4), 1375–1421, 1979.

Horne, J.: Water policy responses to drought in the MDB, Australia, Water Policy Oxf., 18(S2), 28–51, doi:http://dx.doi.org.ezproxy.library.wur.nl/10.2166/wp.2016.012, 2016.

Hudson, D., Marshall, A. G., Yin, Y., Alves, O. and Hendon, H. H.: Improving Intraseasonal Prediction with a New Ensemble Generation Strategy, Mon. Weather Rev., 141(12), 4429–4449, doi:10.1175/MWR-D-13-00059.1, 2013.

van Huijgevoort, M. H. J., Hazenberg, P., van Lanen, H. a. J., Teuling, A. J., Clark, D. B., Folwell, S., Gosling, S. N., Hanasaki, N., Heinke, J., Koirala, S., Stacke, T., Voss, F., Sheffield, J. and Uijlenhoet, R.: Global Multimodel Analysis of Drought in Runoff for the Second Half of the Twentieth Century, J. Hydrometeorol., 14(5), 1535–1552, doi:10.1175/JHM-D-12-0186.1, 2013.

IDEAM: Estudio Nacional del Agua. Instituto de Hidrología, Meteorología y Estudios Ambientales, Colombia, 2015.

IMN: Boletin del ENOS No 74. Instituto Meterológico Nacional, Costa Rica, [online] Available from: https://www.imn.ac.cr/documents/10179/28158/%23%2074, 2014.

Ittersum, M. K. van, Bussel, L. G. J. van, Wolf, J., Grassini, P., Wart, J. van, Guilpart, N., Claessens, L., Groot, H. de, Wiebe, K., Mason-D'Croz, D., Yang, H., Boogaard, H., Oort, P. A. J. van, Loon, M. P. van, Saito, K., Adimo, O., Adjei-Nsiah, S., Agali, A., Bala, A., Chikowo, R., Kaizzi, K., Kouressy, M., Makoi, J. H. J. R., Ouattara, K., Tesfaye, K. and Cassman, K. G.: Can sub-Saharan Africa feed itself?, Proc. Natl. Acad. Sci., 113(52), 14964–14969, doi:10.1073/pnas.1610359113, 2016.

Jiang, C., Xiong, L., Wang, D., Liu, P., Guo, S. and Xu, C.-Y.: Separating the impacts of climate change and human activities on runoff using the Budyko-type equations with time-varying parameters, J. Hydrol., 522, 326–338, doi:10.1016/j.jhydrol.2014.12.060, 2015.

Karimi, P. and Bastiaanssen, W. G. M.: Spatial evapotranspiration, rainfall and land use data in water accounting – Part 1: Review of the accuracy of the remote sensing data, Hydrol Earth Syst Sci, 19(1), 507–532, doi:10.5194/hess-19-507-2015, 2015.

Karimi, P., Bastiaanssen, W. G. M. and Molden, D.: Water Accounting Plus (WA+) – a water accounting procedure for complex river basins based on satellite measurements, Hydrol Earth Syst Sci, 17(7), 2459–2472, doi:10.5194/hess-17-2459-2013, 2013.

Kauffeldt, A., Wetterhall, F., Pappenberger, F., Salamon, P. and Thielen, J.: Technical review of large-scale hydrological models for implementation in operational flood forecasting schemes on continental level, Environ. Model. Softw., 75, 68–76, doi:10.1016/j.envsoft.2015.09.009, 2016.

Kaune, A., Werner, M., Rodríguez, E. and de Fraiture, C.: Constraining uncertainties in water supply reliability in a tropical data scarce basin, Eur. Geosci. Union Gen. Assem., doi:Vol. 17, EGU2015-11871, 2015., 2015.

Kaune, A., Werner, M., Rodríguez, E., Karimi, P. and de Fraiture, C.: A novel tool to assess available hydrological information and the occurrence of sub-optimal water allocation decisions in large irrigation districts, Agric. Water Manag., 191(Supplement C), 229–238, doi:10.1016/j.agwat.2017.06.013, 2017.

Kaune, A., López-López, P., Gevaert, A., Veldkamp, T., Werner, M. and de Fraiture, C.: The benefit of using an ensemble of global hydrological models in surface water availability for irrigation area planning, Water Resour. Manag. Rev., 2018.

Khan, S., Rana, T., Beddek, R., Paydar, Z., Carroll, J. and Blackwell, J.: Whole of catchment water and salt balance to identify potential water saving options in the Murrumbidgee Catchment, [online] Available from: https://publications.csiro.au/rpr/pub?list=BRO&pid=procite:af374813-b17a-401f-b81a-3f59a1cf3584 (Accessed 27 October 2018), 2004.

Khan, S., Tariq, R., Yuanlai, C. and Blackwell, J.: Can irrigation be sustainable?, Agric. Water Manag., 80(1–3), 87–99, doi:10.1016/j.agwat.2005.07.006, 2006.

Kim, D. and Kaluarachchi, J. J.: A risk-based hydro-economic analysis for land and water management in water deficit and salinity affected farming regions, Agric. Water Manag., 166, 111–122, doi:10.1016/j.agwat.2015.12.019, 2016.

King, A.: Technology: The Future of Agriculture, Nature, 544, S21–S23, doi:10.1038/544S21a, 2017.

Kirby, J. M., Mainuddin, M., Ahmad, M. D. and Gao, L.: Simplified Monthly Hydrology and Irrigation Water Use Model to Explore Sustainable Water Management Options in the Murray-Darling Basin, Water Resour. Manag., 27(11), 4083–4097, doi:10.1007/s11269-013-0397-x, 2013.

Kirby, J. M., Connor, J., Ahmad, M. D., Gao, L. and Mainuddin, M.: Climate change and environmental water reallocation in the Murray–Darling Basin: Impacts on flows, diversions and economic returns to irrigation, J. Hydrol., 518, Part A, 120–129, doi:10.1016/j.jhydrol.2014.01.024, 2014a.

Kirby, M., Bark, R., Connor, J., Qureshi, M. E. and Keyworth, S.: Sustainable irrigation: How did irrigated agriculture in Australia's Murray–Darling Basin adapt in the Millennium Drought?, Agric. Water Manag., 145, 154–162, doi:10.1016/j.agwat.2014.02.013, 2014b.

Kirby, M., Connor, J., Ahmad, M. D., Gao, L. and Mainuddin, M.: Irrigator and Environmental Water Management Adaptation to Climate Change and Water Reallocation in the Murray–Darling Basin, Water Econ. Policy, 01(03), 1550009, doi:10.1142/S2382624X15500095, 2015.

Knijff, J. M. V. D., Younis, J. and Roo, A. P. J. D.: LISFLOOD: a GIS-based distributed model for river basin scale water balance and flood simulation, Int. J. Geogr. Inf. Sci., 24(2), 189–212, doi:10.1080/13658810802549154, 2010.

Krinner, G., Viovy, N., de Noblet-Ducoudré, N., Ogée, J., Polcher, J., Friedlingstein, P., Ciais, P., Sitch, S. and Prentice, I. C.: A dynamic global vegetation model for studies of the coupled atmosphere-biosphere system, Glob. Biogeochem. Cycles, 19(1), GB1015, doi:10.1029/2003GB002199, 2005.

Le Quesne, T., Pegram, G. and Von der Heyden, C.: Allocating Scarce Water: A primer on Water allocation, water rights and water markets, [online] Available from: http://assets.wwf.org.uk/downloads/scarce_water.pdf, 2007.

Leeuw, J. de, Methven, J. and Blackburn, M.: Evaluation of ERA-Interim reanalysis precipitation products using England and Wales observations, Q. J. R. Meteorol. Soc., 141(688), 798–806, doi:10.1002/qj.2395, n.d.

Linés, C., Werner, M. and Bastiaanssen, W.: The predictability of reported drought events and impacts in the Ebro Basin using six different remote sensing data sets, Hydrol Earth Syst Sci, 21(9), 4747–4765, doi:10.5194/hess-21-4747-2017, 2017.

Linés, C., Iglesias, A., Garrote, L., Sotés, V. and Werner, M.: Do users benefit from additional information in support of operational drought management decisions in the Ebro basin?, Hydrol. Earth Syst. Sci., 22(11), 5901–5917, doi:https://doi.org/10.5194/hess-22-5901-2018, 2018.

Loon, A. F. V. and Lanen, H. A. J. V.: Making the distinction between water scarcity and drought using an observation-modeling framework, Water Resour. Res., 49(3), 1483–1502, doi:10.1002/wrcr.20147, 2013.

Loon, A. F. V., Stahl, K., Baldassarre, G. D., Clark, J., Rangecroft, S., Wanders, N., Gleeson, T., Dijk, A. I. J. M. V., Tallaksen, L. M., Hannaford, J., Uijlenhoet, R., Teuling, A. J., Hannah, D. M., Sheffield, J., Svoboda, M., Verbeiren, B., Wagener, T. and Lanen, H. A. J. V.: Drought in a human-modified world: reframing drought definitions, understanding, and analysis approaches, Hydrol. Earth Syst. Sci., 20(9), 3631–3650, doi:https://doi.org/10.5194/hess-20-3631-2016, 2016.

Lopez Lopez, P.: Application of global hydrological datasets for river basin modelling, [online] Available from: http://dspace.library.uu.nl/handle/1874/364148 (Accessed 8 March 2019), 2018.

López López, P., Wanders, N., Schellekens, J., Renzullo, L. J., Sutanudjaja, E. H. and Bierkens, M. F. P.: Improved large-scale hydrological modelling through the assimilation of streamflow and downscaled satellite soil moisture observations, Hydrol Earth Syst Sci, 20(7), 3059–3076, doi:10.5194/hess-20-3059-2016, 2016.

López López, P., Sutanudjaja, E. H., Schellekens, J., Sterk, G. and Bierkens, M. F. P.: Calibration of a large-scale hydrological model using satellite-based soil moisture and evapotranspiration products, Hydrol Earth Syst Sci, 21(6), 3125–3144, doi:10.5194/hess-21-3125-2017, 2017.

Luseno, W. K., McPeak, J. G., Barrett, C. B., Little, P. D. and Gebru, G.: Assessing the Value of Climate Forecast Information for Pastoralists: Evidence from Southern Ethiopia and Northern Kenya, World Dev., 31(9), 1477–1494, doi:10.1016/S0305-750X(03)00113-X, 2003.

Lutz, A. F., Immerzeel, W. W., Shrestha, A. B. and Bierkens, M. F. P.: Consistent increase in High Asia's runoff due to increasing glacier melt and precipitation, Nat. Clim. Change, 4(7), 587–592, doi:10.1038/nclimate2237, 2014.

Macauley, M. K.: The value of information: Measuring the contribution of space-derived earth science data to resource management, Space Policy, 22(4), 274–282, doi:10.1016/j.spacepol.2006.08.003, 2006.

Malano, H. M. and Hofwegen, P. van: Management of Irrigation and Drainage Systems, CRC Press, Rotterdam; Brookfield Vt., 1999.

Malings, C. and Pozzi, M.: Value of information for spatially distributed systems: Application to sensor placement, Reliab. Eng. Syst. Saf., 154, 219–233, doi:10.1016/j.ress.2016.05.010, 2016.

Marshall, A. G., Hudson, D., Wheeler, M. C., Alves, O., Hendon, H. H., Pook, M. J. and Risbey, J. S.: Intra-seasonal drivers of extreme heat over Australia in observations and POAMA-2, Clim. Dyn., 43(7–8), 1915–1937, doi:10.1007/s00382-013-2016-1, 2014.

Masafu, C. K., Trigg, M. A., Carter, R. and Howden, N. J. K.: Water availability and agricultural demand: An assessment framework using global datasets in a data scarce catchment, Rokel-Seli River, Sierra Leone, J. Hydrol. Reg. Stud., 8, 222–234, doi:10.1016/j.ejrh.2016.10.001, 2016.

MI: Access and Ordering Rules. MI-Murrumbidgee Irrigation Ltd. Australia, [online] Available from: www.mirrigation.com.au, 2013a.

MI: Rice Growing Rules. MI-Murrumbidgee Irrigation Ltd. Australia, [online] Available from: www.mirrigation.com.au, 2013b.

MI: Annual Compliance Report-MI Murrumbidgee Irrigation Ltd, [online] Available from: www.mirrigation.com.au, 2015a.

MI: MIA System Information. MI-Murrumbidgee Irrigation Ltd. Australia, [online] Available from: www.mirrigation.com.au, 2015b.

MINAE: Plan Nacional de Gestión Integrada de los Recursos Hídricos, [online] Available from: http://www.da.go.cr/politicas/, 2008.

MINAE: Política Hídrica Nacional. Ministerio de Ambiente, Energía y Telecomunicaciones, Rector del Recurso y Sector Hídrico. Costa Rica, 2009.

MINAGRICULTURA and INAT: Manual de Normas Técnicas Básicas para la Realización de Proyectos de Adecuación de Tierras. Ministerio de Agricultura, Colombia, [online] Available from: http://bibliotecadigital.agronet.gov.co/jspui/bitstream/11348/6674/1/180.pdf, 1997.

MinAmbiente: Sistema de información Ambiental de Colombia - SIAC - Política Nacional para la Gestión Integral del Recurso Hídrico, [online] Available from: https://www.siac.gov.co/contenido/contenido.aspx?catID=817&conID=838 (Accessed 28 October 2014), 2014,

Murphy, A. H.: Decision Making and the Value of Forecasts in a Generalized Model of the Cost-Loss Ratio Situation, Mon. Weather Rev., 113(3), 362–369, doi:10.1175/1520-0493(1985)113<0362:DMATVO>2.0.CO;2, 1985.

Naumann, G., Barbosa, P., Garrote, L., Iglesias, A. and Vogt, J.: Exploring drought vulnerability in Africa: an indicator based analysis to be used in early warning systems, Hydrol Earth Syst Sci, 18(5), 1591–1604, doi:10.5194/hess-18-1591-2014, 2014.

Neumann, J. V. and Morgenstern, O.: Theory of Games and Economic Behavior, 3rd ed., Princeton University Press. [online] Available from: http://gen.lib.rus.ec/book/index.php?md5=0500EA03BA90540253F05612C1851D9E (Accessed 18 May 2017), 1966.

Nikouei, A., Zibaei, M. and Ward, F. A.: Incentives to adopt irrigation water saving measures for wetlands preservation: An integrated basin scale analysis, J. Hydrol., 464–465, 216–232, doi:10.1016/j.jhydrol.2012.07.013, 2012.

NSW DPI: Rice growing guide, [online] Available from: https://www.dpi.nsw.gov.au/agriculture/broadacre-crops/summer-crops/rice-development-guides/rice-growing-guide, 2016.

Orth, R., Dutra, E., Trigo, I. F. and Balsamo, G.: Advancing land surface model development with satellite-based Earth observations, Hydrol. Earth Syst. Sci., 21(5), 2483–2495, doi:https://doi.org/10.5194/hess-21-2483-2017, 2017.

Pappenberger, F., Ramos, M. H., Cloke, H. L., Wetterhall, F., Alfieri, L., Bogner, K., Mueller, A. and Salamon, P.: How do I know if my forecasts are better? Using benchmarks in hydrological ensemble prediction, J. Hydrol., 522, 697–713, doi:10.1016/j.jhydrol.2015.01.024, 2015.

Pareeth, S., Karimi, P., Shafiei, M. and De Fraiture, C.: Mapping Agricultural Landuse Patterns from Time Series of Landsat 8 Using Random Forest Based Hierarchial Approach, Remote Sens., 11(5), 601, doi:10.3390/rs11050601, 2019.

Peña-Arancibia, J. L., McVicar, T. R., Paydar, Z., Li, L., Guerschman, J. P., Donohue, R. J., Dutta, D., Podger, G. M., van Dijk, A. I. J. M. and Chiew, F. H. S.: Dynamic identification of summer cropping irrigated areas in a large basin experiencing extreme climatic variability, Remote Sens. Environ., 154, 139–152, doi:10.1016/j.rse.2014.08.016, 2014.

Peña-Arancibia, J. L., Mainuddin, M., Kirby, J. M., Chiew, F. H. S., McVicar, T. R. and Vaze, J.: Assessing irrigated agriculture's surface water and groundwater consumption by combining satellite remote sensing and hydrologic modelling, Sci. Total Environ., 542, Part A, 372–382, doi:10.1016/j.scitotenv.2015.10.086, 2016.

Petheram, C., Bristow, K. L. and Nelson, P. N.: Understanding and managing groundwater and salinity in a tropical conjunctive water use irrigation district, Agric. Water Manag., 95(10), 1167–1179, doi:10.1016/j.agwat.2008.04.016, 2008.

Quiroga, S., Garrote, L., Iglesias, A., Fernández-Haddad, Z., Schlickenrieder, J., de Lama, B., Mosso, C. and Sánchez-Arcilla, A.: The economic value of drought information for water management under climate change: a case study in the Ebro basin, Nat. Hazards Earth Syst. Sci., 11(3), 643–657, doi:10.5194/nhess-11-643-2011, 2011.

Raes, D., Steduto, P., Hsiao, T. C. and Fereres, E.: AquaCrop — The FAO Crop Model to Simulate Yield Response to Water: II. Main Algorithms and Software Description, Agron. J., 101(3), 438–447, doi:10.2134/agronj2008.0140s, 2009.

Raimonet, M., Oudin, L., Thieu, V., Silvestre, M., Vautard, R., Rabouille, C. and Le Moigne, P.: Evaluation of Gridded Meteorological Datasets for Hydrological Modeling, J. Hydrometeorol., 18(11), 3027–3041, doi:10.1175/JHM-D-17-0018.1, 2017.

Restrepo, J. D. and Kettner, A.: Human induced discharge diversion in a tropical delta and its environmental implications: The Patía River, Colombia, J. Hydrol., 424–425, 124–142, doi:10.1016/j.jhydrol.2011.12.037, 2012.

Ribbons, C.: Water availability in New South Wales Murray-Darling Basin regulated rivers, [online] Available from: https://www.industry.nsw.gov.au/__data/assets/pdf_file/0008/153926/water_availability_mdb _reg_rivers.pdf, 2009.

Richter, B.: Chasing Water: A Guide for Moving from Scarcity to Sustainability, Island Press., 2014.

Rijkswaterstaat: Deltawerken, [online] Available from: https://www.rijkswaterstaat.nl/water/waterbeheer/bescherming-tegen-het-water/waterkeringen/deltawerken/index.aspx (Accessed 12 July 2018), 2018.

Rijsberman, F. R.: Water scarcity: Fact or fiction?, Agric. Water Manag., 80(1–3), 5–22, doi:10.1016/j.agwat.2005.07.001, 2006.

Rodriguez, E., Sanchez, I., Duque, N., Lopez, P., Kaune, A., Werner, M. and Arboleda, P.: Combined use of local and global hydrometeorological data with regional and global hydrological models in the Magdalena - Cauca river basin, Colombia, Vienna, Austria. [online] Available from: http://meetingorganizer.copernicus.org/EGU2017/EGU2017-10477.pdf, 2017.

Schellekens, J., Dutra, E., Martínez-de la Torre, A., Balsamo, G., Dijk, A. van, Sperna Weiland, F., Minvielle, M., Calvet, J.-C., Decharme, B., Eisner, S., Fink, G., Flörke, M., Peßenteiner, S., Beek, R. van, Polcher, J., Beck, H., Orth, R., Calton, B., Burke, S., Dorigo, W. and Weedon, G. P.: A global water resources ensemble of hydrological models: the eartH2Observe Tier-1 dataset, Earth Syst. Sci. Data, 9(?), 389–413, doi:https://doi.org/10.5194/essd-9-389-2017, 2017.

Schultz, B., Tardieu, H. and Vidal, A.: Role of water management for global food production and poverty alleviation, Irrig. Drain., 58(S1), S3–S21, doi:10.1002/ird.480, 2009.

Seckler, D., Upali, A., Molden, D., de Silva, R. and Barker, R.: World water demand and supply, 1990 to 2025: Scenarios and issues., 1998.

SENARA: Demandas de Agua en el Distrito de Riego Arenal-Tempisque para el 2014. Servicio Nacional de Aguas Subterráneas, Riego y Avenamiento (SENARA), Costa Rica, 2014.

Shukla, M. K.: Soil Physics: An Introduction, 1 edition., CRC Press, Boca Raton., 2013.

Shukla, S., McNally, A., Husak, G. and Funk, C.: A seasonal agricultural drought forecast system for food-insecure regions of East Africa, Hydrol Earth Syst Sci, 18(10), 3907–3921, doi:10.5194/hess-18-3907-2014, 2014.

Sood, A. and Smakhtin, V.: Global hydrological models: a review, Hydrol. Sci. J., 60(4), 549–565, doi:10.1080/02626667.2014.950580, 2015.

Stahl, K., Tallaksen, L. M., Hannaford, J. and van Lanen, H. A. J.: Filling the white space on maps of European runoff trends: estimates from a multi-model ensemble, Hydrol Earth Syst Sci, 16(7), 2035–2047, doi:10.5194/hess-16-2035-2012, 2012.

Steduto, P., Hsiao, T. C., Raes, D. and Fereres, E.: AquaCrop—The FAO Crop Model to Simulate Yield Response to Water: I. Concepts and Underlying Principles, Agron. J., 101(3), 426, doi:10.2134/agronj2008.0139s, 2009.

Stockdale, T. N., Anderson, D. L. T., Balmaseda, M. A., Doblas-Reyes, F., Ferranti, L., Mogensen, K., Palmer, T. N., Molteni, F. and Vitart, F.: ECMWF seasonal forecast system 3 and its prediction of sea surface temperature, Clim. Dyn., 37(3), 455–471, doi:10.1007/s00382-010-0947-3, 2011.

Svendsen, M.: Irrigation and River Basin Management: Options for Governance and Institutions, Cab Intl, Wallingford, Oxon, UK; Cambridge, MA., 2005.

Tegegne, G., Park, D. K. and Kim, Y.-O.: Comparison of hydrological models for the assessment of water resources in a data-scarce region, the Upper Blue Nile River Basin, J. Hydrol. Reg. Stud., 14, 49–66, doi:10.1016/j.ejrh.2017.10.002, 2017.

Tekleab, S., Uhlenbrook, S., Mohamed, Y., Savenije, H. H. G., Temesgen, M. and Wenninger, J.: Water balance modeling of Upper Blue Nile catchments using a top-down approach, Hydrol Earth Syst Sci, 15(7), 2179–2193, doi:10.5194/hess-15-2179-2011, 2011.

Toté, C., Patricio, D., Boogaard, H., van der Wijngaart, R., Tarnavsky, E. and Funk, C.: Evaluation of Satellite Rainfall Estimates for Drought and Flood Monitoring in Mozambique, Remote Sens., 7(2), 1758–1776, doi:10.3390/rs70201758, 2015.

Turner, M.: Hydrologic Reference Station Selection Guidelines, [online] Available from: http://www.bom.gov.au/water/hrs/media/static/papers/Selection_Guidelines.pdf, 2012.

Turner, S. W. D., Bennett, J. C., Robertson, D. E. and Galelli, S.: Complex relationship between seasonal streamflow forecast skill and value in reservoir operations, Hydrol Earth Syst Sci, 21(9), 4841–4859, doi:10.5194/hess-21-4841-2017, 2017.

Turral, H., Svendsen, M. and Faures, J. M.: Investing in irrigation: Reviewing the past and looking to the future, Agric. Water Manag., 97(4), 551–560, doi:10.1016/j.agwat.2009.07.012, 2010.

Urrutia-Cobo, N.: Sustainable Management After Irrigation System Transfer: PhD: UNESCO-IHE Institute, Delft (eBook) - Taylor & Francis, [online] Available from: http://tandf.net/books/details/9781466518780/ (Accessed 3 February 2015), 2006.

Veldkamp, T. I. E., Wada, Y., de Moel, H., Kummu, M., Eisner, S., Aerts, J. C. J. H. and Ward, P. J.: Changing mechanism of global water scarcity events: Impacts of socioeconomic changes and inter-annual hydro-climatic variability, Glob. Environ. Change, 32(Supplement C), 18–29, doi:10.1016/j.gloenvcha.2015.02.011, 2015a.

Veldkamp, T. I. E., Eisner, S., Wada, Y., Aerts, J. C. J. H. and Ward, P. J.: Sensitivity of water scarcity events to ENSO-driven climate variability at the global scale, Hydrol Earth Syst Sci, 19(10), 4081–4098, doi:10.5194/hess-19-4081-2015, 2015b.

Verkade, J. S. and Werner, M. G. F.: Estimating the benefits of single value and probability forecasting for flood warning, Hydrol Earth Syst Sci Discuss, 8(4), 6639–6681, doi:10.5194/hessd-8-6639-2011, 2011.

Vermillion, D. L. and Garcés-Restrepo, C.: Results of management turnover in two irrigation districts in Colombia, [online] Available from: http://www.iwmi.cgiar.org/Publications/IWMI_Research_Reports/PDF/pub004/REPORT04.PDF (Accessed 25 November 2014), 1996.

Wada, Y., Wisser, D. and Bierkens, M. F. P.: Global modeling of withdrawal, allocation and consumptive use of surface water and groundwater resources, Earth Syst Dynam, 5(1), 15–40, doi:10.5194/esd-5-15-2014, 2014.

Wada, Y., Flörke, M., Hanasaki, N., Eisner, S., Fischer, G., Tramberend, S., Satoh, Y., Vliet, M. T. H. van, Yillia, P., Ringler, C., Burek, P. and Wiberg, D.: Modeling global water use for the 21st century: the Water Futures and Solutions (WFaS) initiative and its approaches, Geosci. Model Dev., 9(1), 175–222, doi:https://doi.org/10.5194/gmd-9-175-2016, 2016.

Weedon, G. P., Balsamo, G., Bellouin, N., Gomes, S., Best, M. J. and Viterbo, P.: The WFDEI meteorological forcing data set: WATCH Forcing Data methodology applied to ERA-Interim reanalysis data, Water Resour. Res., 50(9), 7505–7514, doi:10.1002/2014WR015638, 2014.

Wilks, D. S.: Statistical Methods in the Atmospheric Sciences, Edición: 3., Academic Press, Amsterdam ; Boston., 2011.

Winsemius, H. C., Dutra, E., Engelbrecht, F. A., Van Garder en, E. A., Wetterhall, F., Pappenberger, F. and Werner, M. G. F.: The potential value of seasonal forecasts in a changing climate in southern Africa, Hydrol. Earth Syst. Sci. Katlenburg-Lindau, 18(4), 1525, doi:http://dx.doi.org/10.5194/hess-18-1525-2014, 2014.

WMO: Guide to Hydrological Practices. Volume I. Hydrology - From Measurement to Hydrological Information. World Meteorological Organization (WMO-No. 168). Sixth edition., 2008.

Zhang, L., Potter, N., Hickel, K., Zhang, Y. and Shao, Q.: Water balance modeling over variable time scales based on the Budyko framework – Model development and testing, J. Hydrol., 360(1–4), 117–131, doi:10.1016/j.jhydrol.2008.07.021, 2008.

Zhang, S. X., Bari, M., Amirthanathan, G., Kent, D., MacDonald, A. and Shin, D.: Hydrologic reference stations to monitor climate-driven streamflow variability and trends, Hydrol. Water Resour. Symp. 2014, 1048, 2014.

Zhang, Y., Zheng, H., Chiew, F. H. S., Arancibia, J. P. and Zhou, X.: Evaluating Regional and Global Hydrological Models against Streamflow and Evapotranspiration Measurements, J. Hydrometeorol., 17(3), 995–1010, doi:10.1175/JHM-D-15-0107.1, 2016.

Zhao, F., Veldkamp, T. I. E., Frieler, K., Schewe, J., Ostberg, S., Willner, S., Bernhard Schauberger, Gosling, S. N., Schmied, H. M., Portmann, F. T., Leng, G., Huang, M., Xingcai Liu, Tang, Q., Hanasaki, N., Biemans, H., Gerten, D., Satoh, Y., Pokhrel, Y., Tobias Stacke, Ciais, P., Chang, J., Ducharne, A., Guimberteau, M., Wada, Y., Hyungjun Kim and Yamazaki, D.: The critical role of the routing scheme in simulating peak river discharge in global hydrological models, Environ. Res. Lett., 12(7), 075003, doi:10.1088/1748-9326/aa7250, 2017.

LIST OF ACRONYMS

DRAT: Arenal-Tempisque Irrigation District

MIA: Murrumbidgee Irrigation Area or Murrumbidgee Irrigation District

WMO: World Meteorological Organization

r_i : Period of record for hydro-meteorological variable i

r_j : Period of record in established control locations j

Q: River discharge

D: Irrigation demand

S: Water supply

S_d: Water storage in the reservoirs

r: Irrigation demand rate

U: Expected annual utility

U_c: Cummulative water use due to allocation

U_d: Daily expected allocation for the different water users

R: Risky annual production

R_a: Annual reserve of water storage

RUV: Relative utility value

$PRUV$: Pooled relative utility value

K_y: Crop sensitivity factor

CHIRPS: Climate Hazards Group InfraRed Precipitation with Station data

MSWEP: Multi-Source Weighted-Ensemble Precipitation

ESP: Extended Streamflow Prediction

POAMA: M2.4 seasonal climate forecasting system

FoGSS: Forecast Guided Stochastic Scenarios

WA: Allocated water

AW: Available water for allocation

LIST OF TABLES

LIST OF FIGURES

ABOUT THE AUTHOR

Born in Hamburg, Germany and raised in the tropical climate of Costa Rica. Agricultural Engineer with an MSc degree in Hydraulic Engineering. Ten years of work experience in water and agriculture projects in the commercial, government, and research sector in the Netherlands, Costa Rica, Colombia, Angola, Georgia, Iran and Australia. Experience in consultancy and research projects for hydrological assessment, climate change impact, biomass growth evaluation, and irrigation advice using ground data, numerical prediction tools and remote sensing. Skilled in project management, valuation methods, simulation models, risk assessment and hydraulic design. Passionate about solving problems related to the environment, water allocation, food and fibre production and irrigation system operation and planning.

Journal publications

Kaune, A., Werner, M., Rodríguez, E., Karimi P., de Fraiture C., 2017. A novel tool to assess available hydrological information and the occurrence of sub-optimal water allocation decisions in large irrigation districts. Agric. Water Manag. 191, 229-238. https://doi.org/10.1016/j.agwat.2017.06.013.

Kaune, A., Werner, M., López López, P., Rodríguez, E., Karimi, P., and de Fraiture, C.: Can global precipitation datasets benefit the estimation of the area to be cropped in irrigated agriculture?, Hydrol. Earth Syst. Sci., 23, 2351-2368, https://doi.org/10.5194/hess-23-2351-2019, 2019.

Kaune A., López-López P., Gevaert A., Veldkamp T., Werner M., de Fraiture C., 2018. The benefit of using an ensemble of global hydrological models in surface water availability for irrigation area planning. Water Resour. Manage. Rev., 2018 (In review).

Kaune A., Chowdhury F., Werner M., Bennett J., 2019. The benefit of using an ensemble of seasonsal streamflow forecasts in water allocation decisions. Hydrol. Earth Syst. Sci., 2019 (In review).

Rodríguez E., Sánchez I., Duque N., Arboleda P., Vega C., Zamora D., López P., Kaune A., Werner M., García C., Burke S., 2018. Combined use of local and global hydrometeorological data with hydrological models for water resources management in the Magdalena-Cauca macrobasin-Colombia. Water Resour. Manage. (2019). https://doi.org/10.1007/s11269-019-02236-5

The research described in this thesis was financially supported by the EartH2Observe project. Additional financial support was provided by CONICIT/MICIT (*Programa de Innovación y Capital Humano para la Competitividad, PINN*) Costa Rica.

Netherlands Research School for the
Socio-Economic and Natural Sciences of the Environment

DIPLOMA

For specialised PhD training

The Netherlands Research School for the
Socio-Economic and Natural Sciences of the Environment
(SENSE) declares that

Alexander José
Kaune Schmidt

born on 10 October 1983 in Hamburg, Germany

has successfully fulfilled all requirements of the
Educational Programme of SENSE.

Delft, 27 September 2019

The Chairman of the SENSE board   the SENSE Director of Education

Prof. dr. Martin Wassen   Dr. Ad van Dommelen

The SENSE Research School has been accredited by the Royal Netherlands Academy of Arts and Sciences (KNAW)

KONINKLIJKE NEDERLANDSE
AKADEMIE VAN WETENSCHAPPEN

The SENSE Research School declares that Alexander José Kaune Schmidt has successfully
fulfilled all requirements of the Educational PhD Programme of SENSE with a
work load of 37.9 EC, including the following activities:

SENSE PhD Courses

- Environmental research in context (2015)
- Research in context activity: 'Preparing and co-organizing annual PhD symposium of IHE
  Delft, Institute for Water Education (Delft, 2-3 October 2017)'

Selection of other PhD and Advanced MSc Courses

- Where there is little data: How to estimate design variables in poorly gauged basins, IHE
  Delft (2014)
- Hydrological modelling course 'Delft FEWS', National University in Colombia (2015)

External training at a foreign research institute

- Interviewing farmers, extension officers, and water associations and 'hydrological
  modelling with the HEC-HMS model', National University in Colombia (2015-2016)
- Field work, data and information collection of water allocation decisions, CSIRO,
  Australia (2016)

Management and Didactic Skills Training

- Teaching in the MSc course 'Irrigation and drainage design' in IHE Delft (2014-2015)
- Workshop organizer in Colombia for the EartH2Observe project (2015)
- Supervising MSc student in IHE Delft with thesis entitled 'Assessment of consistency in
  water allocation decision making' (2018)

Oral Presentations

- *Constraining uncertainties in water supply reliability in a tropical data scarce basin.* EGU
  General Assembly, 12-17 April 2015, Vienna, Austria
- *Hydrological information availability index for water allocation decisions in irrigation
  districts.* World Irrigation Forum, 6-8 November 2016, Chiang, Thailand
- *The benefit of using additional hydrological information from earth observations and
  reanalysis data on water allocation decisions in irrigation districts.* EGU General
  Assembly, 23-28 April, 2017, Vienna, Austria
- *The benefit of using an ensemble of global hydrological models in surface water
  availability for irrigation area planning.* EGU General Assembly, 4-13 April 2018, Vienna,
  Austria

SENSE Coordinator PhD Education

Dr. Peter Vermeulen

T - #0103 - 071024 - C184 - 240/170/10 - PB - 9780367429553 - Gloss Lamination